Tribology and Surface Engineering for Industrial Applications

T0321353

Tribology and Surface Engineering for Industrial Applications

Edited by
Catalin I. Pruncu, Amit Aherwar, and Stanislav Gorb

CRC Press is an imprint of the
Taylor & Francis Group, an **informa** business

First edition published 2022
by CRC Press
6000 Broken Sound Parkway NW, Suite 300, Boca Raton, FL 33487-2742

and by CRC Press
2 Park Square, Milton Park, Abingdon, Oxon, OX14 4RN

ISBN: 978-0-367-49394-3 (hbk)
ISBN: 978-0-367-56260-1 (pbk)
ISBN: 978-1-003-09708-2 (ebk)

DOI: 10.1201/9781003097082

Typeset in Times
by codeMantra

Contents

Preface

Tribology is a multidisciplinary science that encompasses mechanical engineering, materials science, surface engineering (surface coating, surface modification, and surface topography analysis), lubricants, and additives chemistry. Earlier, the tribology – rubbing science – was seen as an emergent and very challenging topic because it brings together different aspects as principles of friction, lubrication, and wear. The knowledge of tribology science was applied first about 3500 BC at a basic level for moving stone in which water was used as a lubricant in order to reduce friction and protect the materials surface, finally leading to reduced wear. Nowadays, the science of tribology is used very broadly and covers the field of metal, polymer, ceramics, and a combination of composites with tremendous applications. To further help the research community and industry environment to be informed with the latest advancement, the proposed book will provide an update of all different materials pairs used in tribological contact in a widespread way. Leading researchers from industry, academia, government, and private research institutions across the globe will benefit from knowledge provided by this book, which is highly application-oriented. Moreover, it provides a cutting-edge research from around the globe on the tribology field. Current status, trends, future directions, opportunities, etc., are discussed in detail, making it friendly for young researchers too. Further, this part of the book is focused mainly on providing a provide a systematic and comprehensive account of the recent progress in lubrication and nanobio tribology.

The contents of this book part are spread over eleven chapters that cover:

Chapter 1. introduces the mechanical and sliding wear performance of AA2024 - AlN / Si_3N_4 hybrid alloy composites using preference selection index method.

Chapter 2. discusses the wear properties of UHMWPE by investigating a gas spring, which includes a piston head made from UHMWPE. Furthermore, wear performance of UHMWPE is given for different boundary conditions.

Chapter 3 provided the knowledge and idea about the selection of suitable reinforcement materials and operating conditions for the aluminium alloys and their composite synthesized by spray forming process for the tribological applications.

Chapter 4 presented the ranking analysis and parametric optimization of ZA27-SiC-Gr alloy composites based upon mechanical and sliding wear performance.

Chapter 5 introduces the surface characteristics of the as machined and post processed additive manufactured parts. Surface texture properties and their importance on the tribological behavior of the surfaces were discussed in terms of different AM surfaces.

Chapter 6 discussed and investigate various materials used in wind turbine, failure analysis of wind turbine components and corrosion aspects of wind turbines followed by commonly used corrosion testing methods.

Chapter 7 discusses about the texturing of cutting tools for machining of super alloys, viz. Inconel 718 and Ti6Al4V. Besides, detailed discussions on the LST of cutting tools, also forms a major part of this work.

Chapter 8 introduces the surface engineering techniques and surface characterisation. In Surface engineering technique, the emphasis is laid on metallic coating and metal substrate. While the later class covers the widely used sophisticated techniques for surface characterisation.

Chapter 9 described the details of various surface engineering processes for petroleum and automotive industries followed by advancement in surface coating techniques.

Chapter 10 presents a review of advances in the tribological practice and knowledge of such traditional processes as cutting tools, rolling, drawing, extrusion, forging and sheet metal forming while giving an insight into surface treatments and finishing processes.

Chapter 11 introduces the mechanism of electroless coating process and role of parameters like pH, nature and concentration of second phase particles, temperature, etc., inelectroless Ni-P coatings have been discussed. In addition, different types of electroless coatings and inclusion of second phase particles in composite coatings have also been summarized.

Editors

Dr. Catalin I. Pruncu is a Researcher Fellow in the Design, Manufacturing, and Engineering Management at the University of Strathclyde, Glasgow, UK, with 10 years of research experience in academia and industry. He published more than 100 papers in ISI journals, 3 books, a patent, and other papers at various national and international conferences. Catalin is a Charter and Member of the Institute of Mechanical Engineers (UK), since November 2015. He has experience in prestigious universities (Imperial College London, University of Birmingham, University of Sussex) and industries such as IMI Truflo Marine Ltd. and Spanish Navy. Recently, he was invited as Editor for Special Issue, Wear Behavior of Polymer Composites and Mathematical Modeling and Simulation in Mechanics and Dynamic Systems, MDPI, and. he is also a reviewer for almost 50 ISI journals including: *Measurement, Elsevier, Journal of Materials Research and Technology, Surface and Coatings Technology, Journal of Cleaner Production*, etc. He was involved in organizing different international conferences including the 12th International Conference on New Trends in Fatigue and Fracture, Brasov, Romania, 2012.

Dr. Amit Aherwar is an Assistant Professor at the Department of Mechanical Engineering, Madhav Institute of Technology and Science, Gwalior, Madhya Pradesh, India. He received his PhD from Malaviya National Institute of Technology, Jaipur, Rajasthan, India. He has more than eight years of teaching and research experience. His research interest includes tribology, biomaterials, surface characterization, multi-material and advanced composites, recycle/reuse of industrial wastes for engineering applications, and multi-criteria optimization. He has published more than 35 technical papers in reputed national and international journals/conferences, and also served as a reviewer for various journals.

Prof. Dr. Stanislav Gorb is a Professor and Director at the Zoological Institute of the Kiel University, Germany. He received his PhD degree in zoology and entomology at the Schmalhausen Institute of Zoology of the Ukrainian Academy of Sciences in Kiev (Ukraine). Gorb was a postdoctoral researcher at the University of Vienna (Austria), a research assistant at the University of Jena, and a group leader at the Max Planck Institutes for Developmental Biology in Tübingen and for Metals Research in Stuttgart (Germany). His research focuses on morphology, structure, biomechanics, physiology, and evolution of surface-related functional systems in animals and plants, as well as the development of biologically inspired technological surfaces and systems. He received the Schlossmann Award in Biology and Materials Science in 1995, International Forum Design Gold Award in 2011, and Materialica "Best of" Award in 2011. In 1998, he was the BioFuture Competition winner for his works on biological attachment devices as possible sources for biomimetics. Gorb is a Corresponding Member of Academy of the Science and Literature Mainz, Germany (since 2010) and Member of the National Academy of Sciences Leopoldina, Germany (since 2011). Gorb has authored several books, more than 500 papers in peer-reviewed journals, and four patents.

Contributors

A. Ansari
Department of Physics
Graphic Era deemed to be University
Dehradun, India

L. Arulmani
Department of Mechanical Engineering
RRIT
Bengaluru, India

J. Ashok Raj
Department of Automobile Engineering
Bharath Institute of Higher Education
    and Research: BIHER
Chennai, India

Sourabh Bhaskar
Mechanical Engineering Department
Malaviya National Institute of
    Technology
Jaipur, Rajasthan, India

S. K. Chourasiya
Department of Mechanical Engineering
Madhav Institute of Technology and
    Science
Gwalior, Madhya Pradesh, India

G. Gautam
Department of Metallurgical and
    Materials Engineering
IIT Roorkee
Roorkee, Uttarakhand, India

Selim Gürgen
Eskişehir Vocational School
ESOGU
Eskişehir, Turkey

Jitendra Kumar Katiyar
Department of Mechanical Engineering
SRM Institute of Science and
    Technology
Kattankulathur, Chennai, Tamil Nadu,
    India

Ashiwani Kumar
Mechanical Engineering Department
Feroze Gandhi Institute of Engineering
    and Technology
Rae Bareli, Uttar Pradesh, India

Kuldeep Kumar
Department of Physics
JV Jain College, CCS University
Meerut, India

Mukesh Kumar
Mechanical Engineering Department
Malaviya National Institute of
    Technology
Jaipur, Rajasthan, India

N. Kumar
Department of Mechanical Engineering
BIET
Jhansi, Uttar Pradesh, India

Melih Cemal Kuşhan
Department of Mechanical Engineering
ESOGU
Eskişehir, Turkey

Aayushi Meena
Mechanical Engineering Department
SVNIT
Surat, India

Jyoti Menghani
Mechanical Engineering Department
SVNIT
Surat, India

A. Mohan
Department of Physics
IIT (BHU)
Varanasi, Uttar Pradesh, India

S. Mohan
Department of Metallurgical
    Engineering
IIT (BHU)
Varanasi, Uttar Pradesh, India

Prajal Nandedwalia
Mechanical Engineering Department
SVNIT
Surat, India

Cherian Paul
Department of Mechanical Engineering
Saintgits College of Engineering
Kottayam, Kerala, India

Jibin T. Philip
Department of Mechanical Engineering
National Institute of Technology
    Mizoram
Aizawl, Mizoram, India

Munna Ram
Department of Physics
Graphic Era deemed to be University
Dehradun, India

Saravanan Ramesh
Mechanical Engineering Department
SVNIT
Surat, India

Binnur Sagbas
Mechanical Engineering Department
Yildiz Technical University
Besiktas, Istanbul, Turkey

B. H. Santosh Kumar
Department of Mechanical Engineering
GITAM School of Technology
Bengaluru, India

Abdullah Sert
Department of Mechanical Engineering
ESOGU
Eskişehir, Turkey

Narendra Kumar Shakyawal
Mechanical Engineering Department
Malaviya National Institute of
    Technology
Jaipur, Rajasthan, India

Awanish Kumar Sharma
Department of Physics
Graphic Era deemed to be University
Dehradun, India

Sulaxna Sharma
THDC, Institute of Hydropower
    Engineering and Technology
Tehri, India

Gyan Singh
Gochar Mahavidhalya
Rampur Maniharn, India

Rajat Thombare
Mechanical Engineering Department
SVNIT
Surat, India

D. Tijo
Department of Mechanical Engineering
Saintgits College of Engineering
Kottayam, Kerala, India

1 Mechanical and Sliding Wear Performance of AA2024-AlN/Si$_3$N$_4$ Hybrid Alloy Composites Using Preference Selection Index Method

Mukesh Kumar, Sourabh Bhaskar, and Narendra Kumar Shakyawal
Malaviya National Institute of Technology

Ashiwani Kumar
Feroze Gandhi Institute of Engineering and Technology

CONTENTS

DOI: 10.1201/9781003097082-1

1.1 INTRODUCTION

Aluminum metal is one of the abundantly available metals in our earth's crust. Owing to its peculiar features such as lightweight, manufacturability, mechanical properties, etc., material engineers and scholars are actively working in order to enhance the performance usage as part components across industries. Consequently, various grades of cast and wrought alloys are prepared. The functional features of such alloys could further be enhanced by reinforcing them with suitable ceramic particulates as per application requirements. AA2024 wrought alloy is widely used for tribological applications, and research is still going on to enhance its functional features by mode of alloy composites [1]. Some significant research has been going on with solid lubricant particles like graphite along with hard ceramic particulates to enhance tribological performance [2]. Various studies are available to describe the effects of ceramic reinforcements on the tribological, mechanical, and fracture properties like Bhaskar et al. [3–5] reported AA2024 alloy composite reinforced with SiC, Si_3N_4, and graphite particulates. They observed a significant enhancement in physical, mechanical, thermal, thermomechanical, fracture, and tribological performance. In addition, popular decision-making techniques like the hybrid AHP-TOPSIS technique could be successfully used for ranking the compositions. Kumar and Kumar [6] found significant improvement in the sliding wear performance of AA6061 alloy composite reinforced with graphite particulates. The Taguchi design of the experimental approach is applied and obtained that the sliding velocity is the most influential factor affecting the specific wear rate. Kumar and Kumar [7] found significant improvement in mechanical and sliding wear characteristics of the AA7075-B_4C-rice husk ash alloy composite. Devaraju et al. [8] reported improvement in wear resistance of AA6061-T6- Gr/Al_2O_3-SiC alloy composites. Rao et al. [9] and Rao and Das [10] reported improvement in wear resistance with the addition of SiC particulates in Al-Zn-Mg-Cu alloy. Hong et al. [11] observed improvement in mechanical properties of AA2024 alloy composites reinforced with (3, 5, 7, and 10 vol.%) SiC particles. Selvam et al. [12] observed an increasing trend in mechanical properties of AA6061-($TiB_2 + Al_2O_3$) alloy composites. Sharma et al. [13] reported improved physical and mechanical characteristics of AA6082-Si_3N_4 alloy composites. Both hardness and tensile strength were significantly enhanced with the amount of Si_3N_4 ceramic particulates. Umanath et al. [14] observed higher sliding wear resistance of Al-(SiC + Al_2O_3) alloy composites. Dharmalingam et al. [15] reported improved wear performance of Al-Si-10Mg)-MoS_2/Al_2O_3 alloy composites. Ozdemir et al. [16] observed higher UTS and yield strength of fine particulates relative to coarse particulates in AA2017-SiC/Al_2O_3 alloy composites. Rajeev et al. [17] observed higher wear resistance with silicon particulates in A319 and A390 alloys. Herbert et al. [18] observed higher wear resistance of as-cast Al-Cu-TiB_2 alloy composites. Kurt [19] observed an increase in the UTS of AA5083 when reinforced with Gr, SiC, ZrO_2, CNT, and Al_2O_3.

Stimulated from the literature mentioned above, the novelty of research work presented here comprises (1) design, development, and fabrication of hybrid AA2024-AlN/Si_3N_4 alloy composites with a complementary combination of ceramic particulates through semi-automatic stir casting technique; (2) exploring the synergistic combined effect of both the ceramics on physical, mechanical, thermal conductivity, fracture toughness, and tribological behavior of hybrid alloy composites as per ASTM standards; (3) adopting the Taguchi design of experiment methodology for designing sliding wear experimental simulations as well as input control parameter (such as sliding velocity, sliding distance, normal load, composition, and environment temperature) optimization using Analysis of Variance (ANOVA) and ranking of hybrid alloy composites using preference selection index (PSI) method; and (4) analyzing worn-out surface using SEM/EDXS.

1.2 EXPERIMENTAL DETAILS AND METHODOLOGY

1.2.1 MATERIALS, DESIGN ASPECTS, AND FABRICATION PROCEDURE

The absolute properties of the ingredients of hybrid alloy composites are listed in Table 1.1 [1,20]. The other details such as design aspect and standard industrial fabrication procedure are cataloged in Figure 1.1 [3].

1.2.2 PHYSICAL, MECHANICAL, THERMAL CONDUCTIVITY, AND FRACTURE TOUGHNESS CHARACTERIZATION

The theoretical density (ρ_{ct}) of the alloy composites samples was calculated using Equation 1.1, as proposed by Aggarwal and Broutman [21]:

$$\rho_{ct} = \frac{1}{\dfrac{W_{Al2024}}{\rho_{Al2024}} + \dfrac{W_{SiC}}{\rho_{SiC}} + \dfrac{W_{Si_3N_4}}{\rho_{Si_3N_4}} + \dfrac{W_{Gr}}{\rho_{Gr}}} \tag{1.1}$$

where w and ρ represent the weight fraction and density, respectively. The actual densities (ρ_{ce}) of the specimens were obtained experimentally by a simple water immersion technique following the Archimedes principle, as per ASTM D792. The presence of voids (V_c) in the composite samples was determined using Equation 1.2 [22]:

$$V_c = \frac{\rho_{ct} - \rho_{ce}}{\rho_{ct}} \tag{1.2}$$

TABLE 1.1
Properties of Materials Used in the Formulation (Supplier's Data)

Materials	Melting Point (°C)	Density (g/cc)	Tensile Strength (MPa)	Flexural Strength (MPa)	Young's Modulus (GPa)	Thermal Conductivity (W/mk)
AA2024	500	2.78	185	237	70–80	193
Graphite	3600	2.28	115	50	4.8	24
Aluminum nitride (AlN)	2200	3.26	197–270	320	330	285
Silicon nitride (Si$_3$N$_4$)	1900	3.31	360	679	317	27

Materials:

✓ AA2024 alloy matrix as supplied by Vijay Prakash Gupta & Sons, New Delhi having Copper (4.3-4.5%), Manganese (0.5-0.6%), Magnesium (1.3-1.5%) and Zinc, Nickel, Chromium, Lead, Bismuth (< 0.5%) and confirmed by Batra Metallurgical & Spectro Station, New Delhi [3].

✓ Graphite (Gr) as supplied by Central Drug House Private Limited, New Delhi; particle size of 99 μm; constant 2 wt.%.

✓ Silicon Nitride (Si_3N_4) as supplied by Triveni Chemicals, Gujarat; particle size of 44 μm.

✓ Aluminum Nitride (AlN) as supplied by Triveni Chemicals, Gujarat; particle size of 44 μm.

Design of hybrid AA2024 - AlN/Si_3N_4 alloy composites:

(i)   NAS04 (ratio of AlN to Si_3N_4 is 0 to 4 wt.%; constant 2 wt.% of Gr; rest AA2024 alloy)

(ii) NAS13 (ratio of AlN to Si_3N_4 is 1 to 3 wt.%; constant 2 wt.% of Gr; rest AA2024 alloy)

(iii) NAS22 (ratio of AlN to Si_3N_4 is 2 to 2 wt.%; constant 2 wt.% of Gr; rest AA2024 alloy)

(iv) NAS31 (ratio of AlN to Si_3N_4 is 3 to 1 wt.%; constant 2 wt.% of Gr; rest AA2024 alloy)

(v)   NAS40 (ratio of AlN to Si_3N_4 is 4 to 0 wt.%; constant 2 wt.% of Gr; rest AA2024 alloy)

Fabrication procedure:

The fabrication of hybrid AA2024-AlN/Si_3N_4 alloy composites was performed using in-house semi-automatic stir casting apparatus. The following procedure as per Industrial standard was adopted [3]:

1. **Cleaning and Cutting of AA2024 alloy rods:** The cleaned tiny pieces of AA2024 alloy rods of 22 mm diameter were melted in the graphite crucible using induction furnace as per designed proportion.

2. **Melting of alloy matrix:** The melt was then held at ~ 800°C for 15 minutes; thereafter temperature lowered to 660°C (to mushy zone i.e. between solidus and liquids temperatures of the alloy).

3. **Preheating of ceramic particulates:** The ceramic particulates preheated separately at 700 °C for 3 h were added as per design proportion.

4. **Addition of Magnesium powder:** In-order-to enhances the wettability of reinforcing phases in the molten alloy melt 2 wt.% Mg powder was added.

5. **Addition/mixing of preheated reinforcing phases:** To achieve a uniform mix of reinforcing phase in the molten melt, automatic stirrer (stainless steel; speed ~ 500 rpm; time = 10 min.) was used.

6. **Pouring and solidification of the molten mix:** The mixture was then poured into a permanent cast-iron mould with dimensions $150 \times 90 \times 10$ mm³and allowed it to solidify to room temperature in air for one hour.

7. **Cutting of specimen samples:** The specimen samples were prepared using a wire EDM machine as per ASTM standard dimensions followed by polishing with emery paper, for various physical, mechanical, tribological, etc. characterizations.

FIGURE 1.1   Materials, design aspects, and fabrication procedure.

The tensile strength of the specimens (dimension of $140 \times 12 \times 10$ mm³; span length $= 65$ mm; as per ASTM D 3039-76) was measured on the Universal Testing Machine (UTM) of Heico Company [3]. The same equipment was used for measuring the flexural strength of specimens (dimension of $127 \times 12.5 \times 4$ mm³; span length $= 70$ mm; as per ASTM D 2344-84) [23]. The toughness of the specimens (dimension of $55 \times 10 \times 10$ mm³; notch at $45°$; 2.5 mm depth; as per ASTM D-256) could be measured by impact test in terms of absorbed energy (J) required to fracture it, by Charpy V-notch test method and Charpy Impact tester machine [24]. The Rockwell hardness using B-scale on Krystal Elmec Rockwell hardness tester, meant for aluminum alloys as per ASTM E18 [3], was adopted for hardness measurement.

The thermal conductivity features of the specimens (dimension of 25 mm $\times$ 25 mm $\times$ 2 mm; as per ASTM E-1530 standard) were investigated by a hot disc method on a thermal constant analyzer (Hot Disk TPS 500, Gothenburg, Sweden). It has a thermal conductivity range of 10–200 W/mK; a temperature range of 80°C–140°C; and thermal resistance range of 0.01–0.05 m²K/W [4].

To measure the material propensity to brittle fracture, fracture toughness was performed. This information is required for safe and reliable part design. The plain-strain fracture toughness analysis at the crack length of 0.1, 0.3, 0.5 mm, specimen size of $3.5 \times 13 \times 60$ mm³ (thickness $\times$ width $\times$ height) as per ASTM-E399 standard were performed through single-edge notched tension method on UTM of Heico Company having cross-head velocity 0.5 mm/min. To compute stress intensity factor (K), Equations 1.3 and 1.4 are adopted [25].

$$K = \frac{P}{B\sqrt{W}} \gamma(\beta) \tag{1.3}$$

$$\gamma(\beta) = \frac{\sqrt{2\tan\dfrac{\pi a}{2W}}}{\cos\dfrac{\pi a}{2W}} \left[0.75 + 2.02\left(\frac{a}{W}\right) + 0.37\left(1 - \sin\frac{\pi a}{3W}\right)^3 \right] \tag{1.4}$$

where P is the maximum load (N) at the fracture point, B is the thickness of the specimen (m), W is the width of the specimen (m), and a is the crack length.

1.2.3 DRY SLIDING WEAR TRIBOMETER

Sliding wear tribometer of pin-on-disc type (supplied by DUCOM; ASTM G99) was used to simulate the experiment test runs [26]. Pin specimen dimensions were $25 \times 10 \times 10$ mm³, and counterbody or disc was made of hardened ground steel (EN-32; 72 HRC, surface roughness of 0.6 µRa). The material loss of pin specimen, before and after the test run, was measured using a precision electronic balance with an accuracy of ± 1 mg. The specific wear rate (mm³/Nm) of the specimens could be computed using Equation 1.5:

$$W_s = \frac{\Delta m}{\rho t V_s F} (\text{mm}^3/\text{Nm}) \tag{1.5}$$

where Δm is the mass loss during a test run (g), ρ is the density of the composite (g/mm^3), t is the test duration (s), V_s is the sliding velocity (m/s), and F is the average normal load (N) [6].

1.2.4 PSI METHOD AND TAGUCHI DESIGN OF EXPERIMENT OPTIMIZATION

The PSI technique is a novel and efficient tool that aids in decision-making whenever finite alternatives have finite attributes/criteria of conflicting nature. Nowadays, such techniques have been receiving attention in areas such as economics, military, materials, management, etc. It is easier to understand and involve fewer numerical computations. The algorithm of the PSI methods has been elaborately discussed [27]. Thus, it enables analysts in making more credible, understandable, compelling, or persuasive recommendations.

The Taguchi design of experiment methodology was adapted (1) to identify the input control factors as listed in Table 1.2 that influence the response performance, i.e., specific wear rate of hybrid alloy composites under investigation; (2) to design experimental runs using identified input control factors; and (3) to identify the order of input control/operating factors that actively monitors the response performance. Standard L_{25} orthogonal array was used for designing experimental trials and corresponding S/N ratio, and ANOVA was computed for analysis as per methodology using MINITAB 17 software. Since the objective here is to have a minimum specific wear rate of hybrid alloy composites, hence smaller-the-better S/N ratio characteristic was adopted as per Equation 1.6 [28]:

$$\text{Smaller-the-better characteristic}: \frac{S}{N} = -10\log\frac{1}{n}\sum y^2 \qquad (1.6)$$

where n refers to the number of trials and y refers to the observed data.

1.2.5 SURFACE MORPHOLOGY STUDIES

The surface morphology images from a scanning electron microscope (supplied by FEI Nova Nano SEM 450) were used to study wear mechanisms responsible for

---

TABLE 1.2
Experimental Input Control Factors and Their Levels

	Factor	Level					Unit
		I	II	III	IV	V	
A:	Sliding velocity	0.565	1.132	1.696	2.262	2.825	m/s
B:	Sliding distance	678	1357	2035	2714	3392	m
C:	Reinforcement Composition	NAS04	NAS13	NAS22	NAS31	NAS40	ASN series (wt%)
D:	Normal load	10	20	30	40	50	N
E:	Environment temperature	30	35	40	45	50	°C

surface damage during sliding wear experiments. Simultaneously, energy-dispersive X-ray spectroscopy data were utilized to analyze the elemental composition and its dispersion on the specimen surface [3,29].

1.3 RESULTS AND DISCUSSION

1.3.1 PHYSICAL, MECHANICAL, THERMAL CONDUCTIVITY, AND FRACTURE TOUGHNESS ANALYSIS

Various physical/mechanical features of hybrid AA2024-AlN/ Si_3N_4 alloy composites are plotted in Figures 1.2 and 1.3, while fracture toughness in terms of stress intensity factor and thermal conductivity features is plotted in Figure 1.4.

The density and voids content features of hybrid AA2024-AlN/Si_3N_4 alloy composites are shown in Figure 1.2. The detrimental trend of theoretical density remains in the range ~ 2.7840–2.7884 g/cc due to the detrimental proportion of dense Si_3N_4 harder phase (1580 kg/mm^2) and gradual complementary proportion of AlN less dense phase (1100 kg/mm^2) [30]. The actual density remains in the range ~2.576–2.735 g/cc, which is much lower in magnitude relative to the former. The presence of voids remains in the range of ~1.85%–7.47%. The actual density follows the trend NAS22 > NAS31 > NAS13 > NAS04 > NAS40, while the voids content follows NAS22 < NAS31 < NAS13 < NAS04 < NAS40 across the formulation. It infers that the NAS22 alloy composite through highest actual density and moderate theoretical value gives

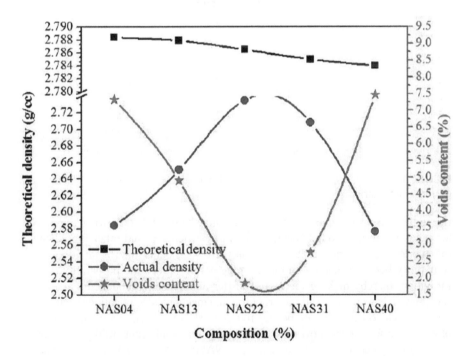

FIGURE 1.2 Density and voids content.

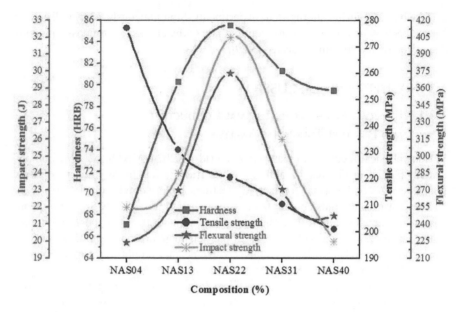

FIGURE 1.3 Mechanical characteristics.

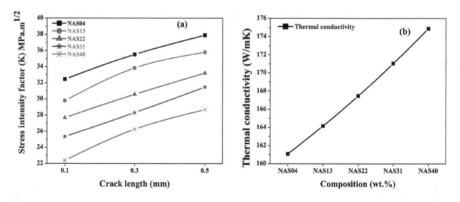

FIGURE 1.4 (a) Stress intensity factor (K) at different crack lengths and (b) thermal conductivity characteristics.

rise to the presence of the lowest voids. This may be possible only when there is better synergy between complementary ceramic particulates and other ingredients of the hybrid alloy composite. It means that there were better interfacial adhesions or wettability between matrix-reinforcing phases that reduce the probability of voids formation, thereby making the matrix stiffer and denser. Components made from such materials would be used for structural and mechanical applications such as gear, liner, etc. The other alloy composites like NAS31/NAS13 with an unequal presence of either reinforcement show next level density and voids content, thereby showing lower synergy between the ingredients. The NAS04/NAS40 alloy composites having either presence of ceramic particulates show last order density and the presence of voids. Such materials have poor interfacial adhesion or wettability between ingredients that promote chances of voids formation-retention that minimizes stiffness of the matrix, making it less dense. Components from such material have lesser reliability under functional structural and mechanical applications [31].

The Rockwell hardness (HRB) of the alloy composites remains within the range ~66.6–85 HRB and follows the trend NAS22 > NAS31 > NAS13 > NAS40 > NAS04. A similar trend was also followed by flexural strength (range ~223.73–373.06 MPa) and impact strength (range ~20–32 J). The tensile strength of the alloy composites is in the range ~201–277.13 MPa and follows the trend NAS04 > NAS13 > NAS22 > NAS31 > NAS40 across the entire formulations. The variation in the trend across the formulation has an in-build impact on the type of ceramic combination as reinforcement in the matrix.

The mechanical characteristics of hybrid AA2024-AlN/Si_3N_4 alloy composites observed to shows a similar trend pattern with slight variation across the formulation. Some significant observations are: (1) NAS22 alloy composite exhibits the presence of the lowest voids that makes the matrix denser and stiffer with better interfacial adhesion strength between the ingredients. Such material when subjected to externally applied mechanical loads/forces (like indenting/flexural/impact/tensile) readily transferred to reinforcing phases that sustained a larger portion leaving minimal for the matrix to sustain. The matrix phase is prevented from any catastrophic failure until all the reinforcing phases fail. This might result in higher mechanical strength magnitudes relative to others. Further, the uniform spread of fine particulate phases into the alloy matrix tends to refine the grain size of the matrix making it stiffer against any kind of deforming or fracturing mechanism [32]. On the other hand, compositions like NAS31/NAS13 alloy composites having unequal presence and NAS40/NAS04 having either presence of reinforcing phases show lower mechanical features. The presence of relatively higher voids might lower the interfacial adhesion strength that makes the load transfer mechanism between matrix-reinforcing phases relatively inefficient. Henceforth, the matrix has to bear considerable load as it retains partial load, along with reinforcements that might be sufficiently high to get deformed or fractured under the application of externally applied mechanical load/force. Further, the voids act as in-build stress raisers and concentrators that add on the external forces in deteriorating the mechanical characteristics of alloy composites [3].

The fracture toughness expressed in terms of stress intensity factor aids in damage tolerance design facet of any product. It is evaluated at different crack lengths such as 0.1, 0.3, and 0.5 mm and is shown in Figure. 1.4a. It is observed that the *"K"* magnitude increases with crack length across the formulation may due to change in the area of cross-section of the crack and increase in the aspect ratio (*a/W*) with crack length. At any crack length, the trend follows NAS04 > NAS13 > NAS22 > NAS31 > NAS40 across the entire formulation. It seems that toughness magnitude decreases with AlN particulates (3.5 MPa.m$^{1/2}$) the presence in complementary combination with Si_3N_4 particulates (8 MPa.m$^{1/2}$). The fewer presence of voids promotes better interfacial adhesion between the ingredients that makes the load transfer mechanism effective between matrix and reinforcement. This delays and minimizes the cycle of crack formation-propagation-deformation-fracture. Further, the hard phase is reported to strengthen the matrix properties by Hall-Petch or grain-boundary strengthening and pinning point mechanism and will improve the fracture toughness of alloy composites [4].

The thermal conductivity analysis of the alloy composites shown in Figure 1.4b shows an incremental trend with the synergistic combined complementary effect of both the ceramic particulates and follows order NAS04 > NAS13 > NAS22 > NAS31 > NAS40 across the formulation. This may be attributed to the incremental presence of AlN particulates with thermal conductivity of 82–170 W/mK relative to the decremented presence of Si_3N_4 particulates (27 W/mK). It seems that reinforcement lowers the overall thermal conductivity features of the alloy matrix. This is probably caused due to the addition of the phase with a lower value than the matrix and the formation of an interface between matrix reinforcement [33]. However, the values are as per industrial standard; hence, the alloy composites could suitably be used for components like rotor disc/drum applications.

1.3.2 ANALYSIS USING PSI METHOD AND TAGUCHI DESIGN OF EXPERIMENT OPTIMIZATION

The stepwise computation of the PSI algorithm is as follows:

Step 1: After brainstorming sessions with experts, the information gathered could be arranged as per the hierarchy structure shown in Figure 1.5. The goal must be at the top, the performance criteria must be at the middle, and material alternatives at the bottom. The identification of performance criteria and their implications are listed in Table 1.3.

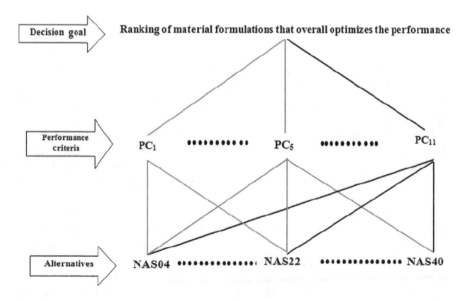

FIGURE 1.5 Hierarchy structure of the investigation problem.

TABLE 1.3

Performance Criteria's and Their Implications [5]

PC's No.	PC's	Implications of PC's
PC-1	Experimental density (g/cc)	Lower the better
PC-2	Voids content (%)	Lower the better
PC-3	Tensile strength (MPa)	Higher the better
PC-4	Flexural strength (MPa)	Higher the better
PC-5	Impact strength (J)	Higher the better
PC-6	Hardness (HRB)	Higher the better
PC-7	Specific Wear Rate (mm^3/Nm)	Lower the better
PC-8	Thermal conductivity (W/mK)	Higher the better
PC-9	Fracture toughness (MPa.$\sqrt{m}$)	Higher the better
PC-10	Coefficient-of-friction	Lower the better
PC-11	Cost (Rs.)	Lower the better

Step 2: The performance parameters as evaluated in Section 1.2 are utilized here as performance criteria, whereas compositions are taken as material alternatives. The decision matrix is as follows (Table 1.4):

Steps 3–7: It involves computation of (1) normalization of the decision matrix, (2) preference variation value (PV$_j$), (3) overall preference value (ψ_j) (here the $\Sigma \psi_j$ comes out to be equal to unity hence consistent), and (4) overall PSI (I_j). Finally, the ranking order of the material alternatives is made based on the descending order of "I_j" score and shown in Table 1.5.

TABLE 1.4

Decision Matrix

		Material Alternatives				
PC's No.	PC's	NAS04	NAS13	NAS22	NAS31	NAS40
PC-1	Experimental density (g/cc)	2.585	2.651	2.735	2.708	2.576
PC-2	Voids content (%)	7.33	4.91	1.85	2.76	7.47
PC-3	Tensile strength (MPa)	277.13	230.91	220.61	210.53	201.01
PC-4	Flexural strength (MPa)	223.73	270.02	373.06	270.75	247.48
PC-5	Impact strength (J)	22	24	32	26	20
PC-6	Hardness (HRB)	66.6	79.8	85	80.8	79
PC-7	Specific wear rate (mm^3/Nm)	5.2100E-05	4.52E-05	3.09E-05	3.95E-05	6.07E-05
PC-8	Thermal conductivity (W/mK)	161.076	164.16	167.47	171.03	174.87
PC-9	Fracture toughness (MPa.$\sqrt{m}$)	37.89	35.79	33.19	31.47	28.68
PC-10	Coefficient of friction	0.21	0.2	0.18	0.19	0.22
PC-11	Cost (Rs.)	512	506	499	493	486

TABLE 1.5
Ranking Order of Material Alternatives

Material Alternatives	NAS04	NAS13	NAS22	NAS31	NAS40
Overall preference index (I_j)	−7.39	−3.77	0.89	−0.63	−7.66
Preference Ranking	4	3	1	2	5

TABLE 1.6
Taguchi's Factorial Experiment Design Using L_{25} Orthogonal Array

S. No.	Sliding Velocity (m/s)	Reinforcement Content (wt%)	Siding Distance (m)	Normal Load (N)	Environment Temperature (°C)	Specific Wear Rate (mm³/Nm)	S/N Ratio (dB)
1.	0.565	NAS04	678	10	30	0.0001895790	74.44
2.	0.565	NAS13	1357	20	35	0.0001381720	77.19
3.	0.565	NAS22	2035	30	40	0.0000415912	87.62
4.	0.565	NAS31	2714	40	45	0.0000451544	86.91
5.	0.565	NAS40	3392	50	50	0.0000073336	102.69
6.	1.132	NAS04	1357	30	45	0.0000947198	80.47
7.	1.132	NAS13	2035	40	50	0.0001126130	78.97
8.	1.132	NAS22	2714	50	30	0.0000218300	93.22
9.	1.132	NAS31	3392	10	35	0.0000722577	82.82
10.	1.132	NAS40	678	20	40	0.0007949440	61.99
11.	1.696	NAS04	2035	50	35	0.0000589512	84.59
12.	1.696	NAS13	2714	10	40	0.0001995820	74.00
13.	1.696	NAS22	3392	20	45	0.0000810949	81.82
14.	1.696	NAS31	678	30	50	0.0005623360	65.00
15.	1.696	NAS40	1357	40	30	0.0001451230	76.77
16.	2.262	NAS04	2714	20	50	0.0001262930	77.97
17.	2.262	NAS13	3392	30	30	0.0000777974	82.18
18.	2.262	NAS22	678	40	35	0.0002808790	71.03
19.	2.262	NAS31	1357	50	40	0.0001685760	75.46
20.	2.262	NAS40	2035	10	45	0.0003259700	69.74
21.	2.825	NAS04	3392	40	40	0.0001010490	79.91
22.	2.825	NAS13	678	50	45	0.0004670600	66.61
23.	2.825	NAS22	1357	10	50	0.0008108300	61.82
24.	2.825	NAS31	2035	20	30	0.0001505520	76.45
25.	2.825	NAS40	2714	30	35	0.0001171170	78.63

The consistent PSI ranking analysis of performance data found to be in line with our subjective analysis i.e., NAS22 > NAS31 > NAS13>NAS04>NAS40 discussed in Section 1.3.1.

Taguchi's factorial design of experiment modus operandi, as discussed in Section 1.2.4, was implemented here for the analysis. Table 1.6 shows the scheme of the experimental simulation, and Figure 1.6 shows the main factor plot for the *S/N* ratio.

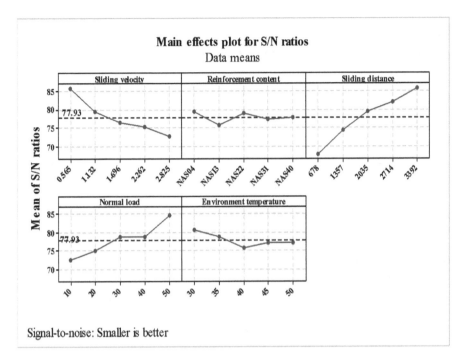

FIGURE 1.6 Main effect plot of input control factors (*S/N* ratio) on SWR.

TABLE 1.7

Ranking Order of Significance of Input Control Factors

Level	A: Sliding Velocity (m/s)	B: Reinforcement Content (wt%)	C: Sliding Distance (m)	D: Normal Load (N)	E: Environment Temperature (°C)
1	85.77	79.48	67.82	72.56	80.61
2	79.49	75.79	74.34	75.08	78.85
3	76.43	79.10	79.47	78.78	75.80
4	75.28	77.33	82.14	78.72	77.11
5	72.68	77.96	85.89	84.52	77.29
Delta	13.09	3.69	18.07	11.95	4.81
Rank	2	5	1	3	4

Table 1.7 shows the ranking order of input control factors, and ANOVA results are shown in Table 1.8. The result outcome from the confirmation experiment is shown in Table 1.9. Minitab 17 software is used for methodology-based computations.

The mean *S/N* ratio for SWR was computed to be 77.93 dB. The ranking order based on the significance of input control factors is sliding distance >> sliding velocity > normal load > environment temperature > reinforcement content and the input control factor combination setting A_1 B_1 C_4 D_4 E_1 (Figure 1.6), i.e., sliding velocity = 0.565 m/s, reinforcement content of NAS04, sliding distance = 3392 m,

TABLE 1.8
ANOVA Results for the Specific Wear Rate

	Input Control Factors	DF	Seq. SS	Adj. SS	Adj. MS	F	p	P (%)
A:	Sliding velocity (m/s)	4	503.67	503.67	125.92	6.32	0.051	24.02
B:	Reinforcement content (wt%)	4	43.56	43.56	10.89	0.55	0.713	2.08
C:	Sliding distance (m)	4	992.94	992.94	248.24	12.47	0.016	47.36
D:	Normal load (N)	4	407.99	407.99	102.00	5.12	0.071	19.46
E:	Environment temperature (°C)	4	68.35	68.35	17.09	0.86	0.557	3.26
	Error	4	79.65	79.65	19.91			
	Total	24	2096.67					

S (4.463); R-sq (96.2%); R-sq (adj) (77.2%); R-sq(pred)(0.00%).
DF, degree of freedom; Seq. SS, sequential sum of squares; Adj. SS, Adjacent sum of square, Adj MS, Adjacent sum of mean square; F, Variance; p, Test statistics; P (%), Percentage contribution of each factor in overall performance (SWR).

TABLE 1.9
Results of the Confirmation Experiment for SWR

	Optimal Control Parameter		
	Prediction	Experimental	Error
Level	$A_2 B_3 C_4 D_4 E_2$	$A_2 B_3 C_4 D_4 E_2$	
S/N ratio for SWR (dB)	85.02	82.15	3.49%

normal load $= 50\,$N and environment temperature $= 30°$C; observed to have optimal SWR. The ANOVA tool could aid to know the extent of contribution of each factor in monitoring response performance of a hybrid alloy composite system under investigation. In this case, the level of significance (denoted by p) of 5% (i.e., the level of confidence of 95%) is considered for the analysis; thus, the lower magnitude of p signifies a higher contribution of that factor in the output response or percentage contribution ($P\%$) may be calculated to understand the same fact. Henceforth, the order of influence of input control factors and their contribution in the SWR is sliding distance ($P = 47.36\%$) $\gg$ sliding velocity ($P = 24.02\%$) > normal load ($P = 19.46\%$) > environment temperature ($P = 3.26\%$) > reinforcement content ($P = 2.08\%$), which is found to be in line with Table 1.8.

The confirmation experimental test was performed to validate the conclusion drawn from the analysis phase, in the last and final stage. For that, consider an arbitrary factor combination setting like $A_2 B_3 C_4 D_4 E_2$ and estimated the S/N ratio using predictive Equation 1.7:

$$\overline{\eta}_1 = \overline{T} + \left[\overline{A}_2 - \overline{T}\right] + \left[\overline{B}_3 - \overline{T}\right] + \left[\overline{C}_4 - \overline{T}\right] + \left[\overline{D}_4 - \overline{T}\right] + \left[\overline{E}_2 - \overline{T}\right] \qquad (1.7)$$

where $\bar{\eta}_1$ is the estimated average; T is the overall experimental average; $\bar{A}_2$, $\bar{B}_3$, $\bar{C}_4$, $\bar{D}_4$, and $\bar{E}_2$ are the average response for the factors at indicated levels. The *S/N* ratio was computed to be $\bar{\eta}_1 = 85.02\,\text{dB}$. This was followed by a confirmation test for the same factor combination experimentally and was found to show 82.15 dB *S/N* ratios. Comparatively, an error of 3.49% is observed, which can further be reduced by increasing experimental runs. In addition, it indicates that the model could be suitably used to predict the SWR of hybrid alloy composites understudy with good accuracy based on input information. Hence, the levels of factors for both steady-state condition and the Taguchi design of experiment are in-tube with each other.

1.3.3 STEADY-STATE-SPECIFIC WEAR RATE AND COEFFICIENT-OF-FRICTION ANALYSIS

The steady-state-specific wear rate (i.e., sliding distance = 678–3392 m; sliding velocity = 1.132 m/s; normal load = 20 N; and environment temperature = 35°C) and coefficient-of-friction of AA2024-AlN/Si$_3$N$_4$ alloy composites are shown in Figure 1.7. It infers that the SWR (Figure 1.7a) and coefficient of friction (Figure 1.7b) of the alloy composites tend to lower with an increase in the sliding distance in general irrespective of the composition; however, sliding test at a certain specific distance results in the following order of wear performance NAS22 > NAS31 > NAS13 > NAS04 > NAS40, while the coefficient of friction follows the order NAS22 < NAS31 < NAS13 < NAS04 < NAS40 across the formulation. During initial lower running cycles, the sharp contact asperities across pin-disc contact surface undergo a repeated cycle of plastic deformation-destruction, thereby generating wear debris. Such debris actively accelerates three-body abrasion across the mating surfaces and generates more debris; this along with debris particles generated through destruction of the tribo-layer accelerates specific wear rates and coefficient of friction of the alloy composites. The compatible synergistic combination between reinforcement-matrix phases results in interfacial adhesion that might be responsible for functional mechanical properties of alloy composites (discussed in Section 1.3.1) that directly or indirectly control the order of performance of SWR and COF. As the sliding distance increases, the

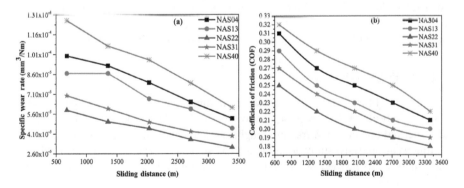

FIGURE 1.7 (a and b) Effect of sliding distance on specific wear rate and coefficient of friction.

contact asperities smoothen out slowly, and due to the generation of sufficient high interfacial flash temperature across the mating surfaces, the mechanism of tribo-film, i.e., formation-deformation-destruction, strengthens due to graphite as a solid lubricant, resulting in minimizing the SWR and frictional forces that might lower the COF of the alloy composites [34].

1.3.4 WORN SURFACE MICROGRAPH ANALYSIS

The damaged surface of sliding wear has been studied using SEM in order to under-stand hidden wear mechanisms responsible for wear rates of the investigated hybrid alloy composites. Henceforth, micrographs (shown in Figure 1.8) of the NAS22 hybrid alloy composite under steady-state conditions are taken for analysis. The micrograph in Figure 1.8a shows severe wear caused by surface destruction, shallow grooves, and deep ploughing action. This is owing to higher contact asperities con-tacts-deformation-rupture phenomenon during the initial running period leading to a higher amount of debris, which further accelerated through a three-body abrasive wear mechanism [3]. The micrographs shown in Figure 1.8b–e indicate relatively lower order of surface damaged, this owing to smoothening out the contact asperi-ties that reduced apparent contact area and formation of tribo-film (due to higher interfacial flash temperature at pin-disc contact, the solid lubricant graphite actively form the lubricating film whose thickness varies with the amount of sliding) reduces metal-to-metal contact, thereby reducing the rate of surface damage during sliding.

The spectra curves and mapping of EDAX analysis (shown in Figure 1.9) of AA2024-AlN/Si_3N_4 alloy composites determine the presence/dispersion of different elements in the alloy composite samples. It shows the presence of elements such as Al, C, O, Cu, Mg, Si, Mn, Fe, Zn, N, Ti, Cr, etc. Al-element presents in the major amount as shown by the largest peak owing to its presence in the alloy and AlN. Similarly, the presence of C-element is due to graphite; Fe (iron) element may be due to counter-surface made of steel; Mg-element may be due to its presence in the alloy and external addition to enhance wettability during fabrication; O-element may be due to the for-mation of oxide layer during sliding action as it was done in the ambient environment; N-element may be due to the presence in Si_3N_4 and AlN particulates; rest elements such as Cr, Cu, Mn, Ti, Zn, Pb presence may be due to alloy composition. The propor-tion of Si-element decreases with the decreasing weight content of Si_3N_4 particulates, but the presence of N-element remains almost constant across the formulation.

1.4 CONCLUSIONS

The novelty of the research work presented in this chapter lies in the design and fabri-cation of hybrid AA2024-AlN/Si_3N_4 alloy composites. This follows physical, mechan-ical, thermal conductivity, fracture toughness, and tribological behavior (steady-state sliding wear analysis) study adopting ASTM standards. The performance data were then analyzed using the PSI method. The significant findings are as follows:

1. Theoretical density remains in the range ~2.7840–2.7884 g/cc and actual density remain in the range ~2.576–2.735 g/cc, while the presence of

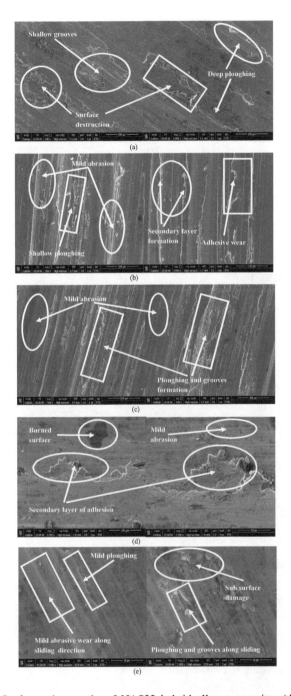

FIGURE 1.8 Surface micrographs of NAS22 hybrid alloy composites (the steady-state-specific wear rate (i.e., sliding distance = 678–3392 m; sliding velocity = 1.132 m/s; normal load = 20 N; and environment temperature = 35°C)). (a) NAS22 at 678 m, (b) NAS22 at 1357 m, (c) NAS22 at 2035 m, (d) NAS22 at 2714 m, and (e) NAS22 at 3392 m.

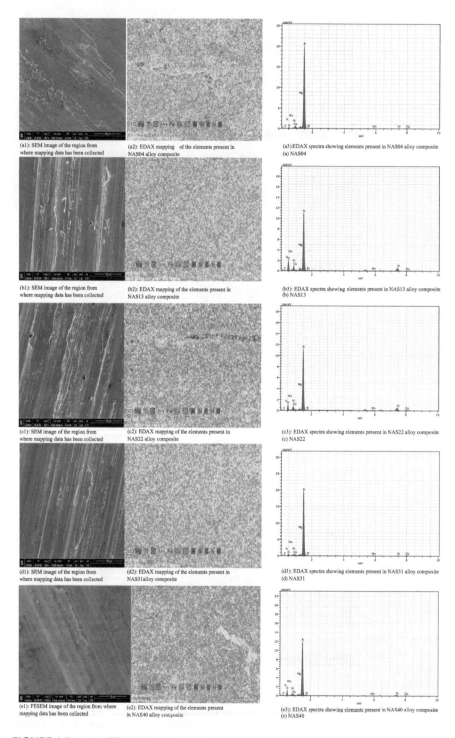

FIGURE 1.9   (a–e) EDAX characteristics.

the voids remains in the range ~1.85%–7.47%. The actual density follows the trend NAS22 > NAS31 > NAS13 > NAS04 > NAS40, while the voids content follows NAS22 < NAS31 < NAS13 < NAS04 < NAS40 across the formulation.

2. The Rockwell hardness (HRB) of the hybrid alloy composites remains within the range ~66.6–85 HRB and follows the trend NAS22 > NAS31 > NAS13 > NAS40 > NAS04. A similar trend was also followed by flexural strength (range ~ 223.73–373.06 MPa) and Impact strength (range ~ 20–32 J). The tensile strength of the alloy composites is in the range ~ 201–277.13 MPa and follows the trend NAS04 > NAS13 > NAS22 > NAS31 > NAS40 across the entire formulations. The NAS22 alloy composites exhibit superior mechanical features relative to others.

3. The fracture toughness expressed as stress intensity factor K increases with crack length (0.1, 0.3, and 0.5 mm). For any crack length, the order of K followed is NAS04 > NAS13 > NAS22 > NAS31 > NAS40 across the entire formulation. The magnitude of K remains within 22–39 MPa.$\sqrt{m}$.

4. The thermal conductivity analysis of the alloy composites shows an incremental trend with the synergistic combined complementary effect of both the ceramic particulates and follows order NAS04 > NAS13 > NAS22 > NAS31 > NAS40 across the formulation.

5. The consistent PSI ranking analysis of performance data found to be in line with our subjective analysis, i.e., NAS22 > NAS31 > NAS13 > NAS04 > NAS40. Thus, MCDM methods like the PSI method could effectively be used for taking decisions whenever there are finite material alternatives and finite performance evaluating criteria that too of conflicting nature.

6. Taguchi's factorial design of the experiment shows the mean S/N ratio for SWR to be 77.93 dB. The ranking order based on the significance of input control factors are sliding distance >> sliding velocity > normal load > environment temperature > reinforcement content and validated by ANOVA results. The confirmation test reveals an error of 3.49% that validates the robustness of the experimental design. The levels of factors for both steady-state condition and the Taguchi design of experiment are in-tube with each other.

7. The steady-state wear and friction analysis infer that both SWR and coefficient of friction of the alloy composites tend to lower with an increase in the sliding distance in general irrespective of the composition; however, sliding test at certain specific distance results in the following order of wear performance NAS22 > NAS31 > NAS13 > NAS04 > NAS40, while the coefficient-of-friction follows the order NAS22 < NAS31 < NAS13 < NAS04 < NAS40 across the formulation. This finding correlated with the mechanical characteristics of the alloy composites.

8. The surface micrographs reveal the wear mechanisms/sub-mechanisms liable for surface damage during sliding wear. The spectra curves and mapping of EDAX analysis reveal the presence/dispersion of different elements in the alloy composite samples.

ACKNOWLEDGMENTS

The authors express their sincere gratitude to the Department of Mechanical Engineering of Malaviya National Institute of Technology, Jaipur-302017, Rajasthan, India for all kinds of financial as well as other miscellaneous infrastructural support. The authors also acknowledge the aid and facilities provided by the Advanced Research Lab for Tribology and Material Research Centre of the Institute for experimentation and characterization work.

REFERENCES

1. E. L. Rooy, *Introduction to Aluminum and Aluminum Alloys: ASM Handbook Properties and Selection: Non Ferrous Alloys and Special Purpose Materials*, Vol. 2, ASM International, United State of America, 1990.
2. S. Bhaskar, M. Kumar and A. Patnaik, A review on tribological and mechanical properties of Al alloy composites, *Mater. Today Proc.* 25 (2019), pp. 810–815.
3. S. Bhaskar, M. Kumar and A. Patnaik, Silicon carbide ceramic particulate reinforced AA2024 alloy composite - Part I: Evaluation of mechanical and sliding tribology performance, *Silicon.* 12 (2019), pp. 843–865.
4. S. Bhaskar, M. Kumar and A. Patnaik, Microstructure, thermal, thermo-mechanical and fracture analyses of hybrid AA2024-SiC alloy composites, *Trans. Indian Inst. Met.* 73 (2019), pp. 181–190.
5. S. Bhaskar, M. Kumar and A. Patnaik, Application of hybrid AHP-TOPSIS technique in analyzing material performance of silicon carbide ceramic particulate reinforced AA2024 alloy composite, *Silicon.* 12 (2019), pp. 1075–1084.
6. M. Kumar and A. Kumar, Sliding wear performance of graphite reinforced AA6061 alloy composites for rotor drum/disk application, *Mater. Today Proc.* 27 (2019), pp. 1972–1976.
7. A. Kumar and M. Kumar, Mechanical and dry sliding wear behaviour of B4C and rice husk ash reinfroced Al 7075 alloy hybrid composite for armors application by using taguchi techniques, *Mater. Today Proc.* 27 (2019), pp. 2617–2625.
8. A. Devaraju, A. Kumar and B. Kotiveerachari, Influence of addition of Grp/Al$_2$O$_3$p with SiCp on wear properties of aluminum alloy 6061-T6 hybrid composites via friction stir processing, *Trans. Nonferrous Met. Soc. China (English Ed)*. 23 (2013), pp. 1275–1280.
9. R.N. Rao, S. Das, D.P. Mondal and G. Dixit, Mechanism of material removal during tribological behaviour of aluminium matrix (Al-Zn-Mg-Cu) composites, *Tribol. Int.* 53 (2012), pp. 179–184.
10. R.N. Rao and S. Das, Effect of SiC content and sliding speed on the wear behaviour of aluminium matrix composites, *Mater. Des.* 32 (2011), pp. 1066–1071.
11. S.J. Hong, H.M. Kim, D. Huh, C. Suryanarayana and B.S. Chun, Effect of clustering on the mechanical properties of SiC particulate-reinforced aluminum alloy 2024 metal matrix composites, *Mater. Sci. Eng. A* 347 (2003), pp. 198–204.
12. J. David Raja Selvam, I. Dinaharan, S. Vibin Philip and P.M. Mashinini, Microstructure and mechanical characterization of in situ synthesized AA6061/(TiB$_2$+Al$_2$O$_3$) hybrid aluminum matrix composites, *J. Alloys Compd.* 740 (2018), pp. 529–535.
13. P. Sharma, S. Sharma and D. Khanduja, Production and some properties of Si$_3$N$_4$ reinforced aluminium alloy composites, *J. Asian Ceram. Soc.* 3 (2015), pp. 352–359.
14. K. Umanath, K. Palanikumar and S.T. Selvamani, Analysis of dry sliding wear behaviour of Al6061/SiC/Al$_2$O$_3$ hybrid metal matrix composites, *Compos. Part B Eng.* 53 (2013), pp. 159–168.

15. S. Dharmalingam, R. Subramanian, K. Somasundara Vinoth and B. Anandavel, Optimization of tribological properties in aluminum hybrid metal matrix composites using gray-taguchi method, *J. Mater. Eng. Perform.* 20 (2011), pp. 1457–1466.
16. I. Ozdemir, S. Muecklich, H. Podlesak and B. Wielage, Thixoforming of AA 2017 aluminum alloy composites, *J. Mater. Process. Technol.* 211 (2011), pp. 1260–1267.
17. V.R. Rajeev, D.K. Dwivedi and S.C. Jain, Dry reciprocating wear of Al-Si-SiCp composites: A statistical analysis, *Tribol. Int.* 43 (2010), pp. 1532–1541.
18. M.A. Herbert, R. Maiti, R. Mitra and M. Chakraborty, Wear behaviour of cast and mushy state rolled Al-4.5Cu alloy and in-situ Al4.5Cu-5TiB$_2$ composite, *Wear* 265 (2008), pp. 1606–1618.
19. H.I. Kurt, Influence of hybrid ratio and friction stir processing parameters on ultimate tensile strength of 5083 aluminum matrix hybrid composites, *Compos. Part B Eng.* 93 (2016), pp. 26–34.
20. *Material Properties Charts*, CoorsTek, Inc., Golden, CO, www.coorstek.com/1999.
21. A. Patnaik, P. Kumar, S. Biswas and M. Kumar, Investigations on micro-mechanical and thermal characteristics of glass fiber reinforced epoxy based binary composite structure using finite element method, *Comput. Mater. Sci.* 62 (2012), pp. 142–151.
22. A. Kumar, A. Patnaik and I.K. Bhat, Effect of titanium metal powder on thermo-mechanical and sliding wear behavior of Al7075/Ti alloy composites for gear application, *Mater. Today Proc.* 5 (2018), pp. 16919–16927.
23. A. Baradeswaran and A. Elaya Perumal, Study on mechanical and wear properties of Al 7075/Al$_2$O$_3$/graphite hybrid composites, *Compos. Part B Eng.* 56 (2014), pp. 464–471.
24. M. Bahrami, N. Helmi, K. Dehghani and M.K.B. Givi, Exploring the effects of SiC reinforcement incorporation on mechanical properties of friction stir welded 7075 aluminum alloy: Fatigue life, impact energy, tensile strength, *Mater. Sci. Eng. A* 595 (2014), pp. 173–178.
25. T. Mamatha, Thermo-mechanical, fracture and erosive wear analysis of particulate filled hybrid metal alloy composites, *A thesis submitted for the award of degree doctor of philosophy in mechanical engineering*, NIT Hamirpur, (2012).
26. A. Kumar, A. Patnaik and I.K. Bhat, Investigation of nickel metal powder on tribological and mechanical properties of Al-7075 alloy composites for gear materials, *Powder Metall.* 60 (2017), pp. 371–383.
27. K. Maniya and M.G. Bhatt, A selection of material using a novel type decision-making method: Preference selection index method, *Mater. Des.* 31 (2010), pp. 1785–1789.
28. A. Kumar, A. Patnaik and I.K. Bhat, Tribology analysis of cobalt particulate filled Al 7075 alloy for gear materials: A comparative study, *Silicon* (2018), pp. 1–17.
29. Information on the FESEM, Radboud University Nijmegen. www.sem.com/analytic/sem.html.
30. N.K. Shakyawal, Sliding wear behavior of Aluminium alloy 2024 composites reinforced with ceramic particulates, *M.Tech. thesis*, MNIT Jaipur, 2018.
31. N. Mathan Kumar, S. Senthil Kumaran and L.A. Kumaraswamidhas, Aerospace application on Al 2618 with reinforced – Si$_3$N$_4$, AlN and ZrB$_2$ in-situ composites, *J. Alloys Compd.* 672 (2016), pp. 238–250.
32. H. Arik, Effect of mechanical alloying process on mechanical properties of α-Si$_3$N$_4$ reinforced aluminum-based composite materials, *Mater. Des.* 29 (2008), pp. 1856–1861.
33. J.M. Molina, M. Rhême, J. Carron and L. Weber, Thermal conductivity of aluminum matrix composites reinforced with mixtures of diamond and SiC particles, *Scr. Mater.* 58 (2008), pp. 393–396.
34. L. Jinfeng, J. Longtao, W. Gaohui, T. Shoufu and C. Guoqin, Effect of graphite particle reinforcement on dry sliding wear of SiC/Gr/Al composites, *Rare Met. Mater. Eng.* 38 (2009), pp. 1894–1898.

2 Wear Properties of UHMWPE

A Case Study of Gas Spring

Selim Gürgen, Abdullah Sert,
and Melih Cemal Kuşhan
Eskişehir Osmangazi University

CONTENTS

2.1   INTRODUCTION

Ultra-high molecular weight polyethylene (UHMWPE) is one of the advanced engineering polymers with excellent physical and mechanical properties. Self-lubrication, wear resistance, anti-impact behavior, chemical inertness, and ecofriendly properties make this polymer a good candidate for various engineering applications. Most of UHMWPE utilization in the world is for industrial applications such as sliders, liners, bearings, and seals in machinery as well as bumpers and sidings for ships and harbors, in which wear and impact resistances are prominent in service conditions. In addition to industrial applications, UHMWPE has been extensively utilized as artificial joint implants in orthopedics due to its self-lubricating and anti-wear properties as well as its biocompatible behavior (Kurtz 2004). Artificial ice rinks are also made from this polymer, which seems like ice with its white and bright appearance. Figure 2.1 shows various industrial products made from UHMWPE.

Polymers are formed by molecular chains unlike crystalline materials such as metals. Polyethylene is produced from ethylene (C_2H_4) gas having a molecular weight of 28. Owing to the polymerization process, the chemical formula of polyethylene becomes $-(C_2H_4)_n-$, where n is the degree of polymerization. UHMWPE may include a large number of repeated ethylene units more than 200,000. Polyethylene can be divided into three main groups in terms of molecular weight such as low-density polyethylene (LDPE), high-density polyethylene (HDPE), and ultra-high molecular

DOI: 10.1201/9781003097082-2

23

FIGURE 2.1   Various industrial products made from UHMWPE.

weight polyethylene (UHMWPE). LDPE generally includes lower molecular weights less than 50,000 g/mol, whereas the molecular weight of HDPE may reach 200,000 g/mol (Kurtz 2004). On the other hand, UHMWPE includes a molecular weight as high as 10 million g/mol (A. Huang, Su, and Liu 2013). UHMWPE has a linear molecular architecture, which consists of long polyethylene chains aligning in the same direction. However, in industrial applications, UHMWPE can be copolymerized with different monomers such as polypropylene to tune the mechanical properties, and therefore, branched-chain structures can be observed in these polymers. At the microstructural level, linear molecular chains are tangled by extending to very long distances, and these structures are stimulated by temperature. Below the melting point, molecular chains are more prone to coil up the C–C bonds, thereby generating chain folds in the microstructure. These chain folds generate locally ordered regions, namely crystalline lamella, in the microstructure. In addition to crystalline lamella, amorphous regions occupy considerably large areas, and the connections of these two distinctive regions are established by tie molecules in the microstructure. Figure 2.2 shows a schematic of molecular chains in UHMWPE. Various factors such as molecular weight, manufacturing conditions (temperature, pressure, etc.), and environmental conditions (loading, etc.) are determinants on the degree of crystallinity. For polymer properties, there are three main temperatures such as glass transition temperature (T_g), melting temperature (T_m), and flow temperature (T_f). Below the glass transition temperature, polymers have brittle molecular chains like glass because insufficient thermal energy hinders the motion of molecular chains, and therefore, upon loading, the response of polymers becomes stretching of molecular chains, which may end up with rupture. However, by increasing temperature above the glass transition temperature, molecular chains in amorphous regions gain mobility and the polymers exhibit melting. The degree of crystallinity affects the melting temperature of polymers because crystalline regions are prone to melt at higher temperatures. At elevated temperatures above the melting point, polymers may turn into a liquid state by a flow transition. This point is called flow temperature; however, high molecular weight polymers are generally prevented from flowing due to the entanglement of long molecular chains within the polymer melt. For example, polyethylene with a molecular weight lower than 500,000 g/mol exhibits a flow transition, whereas it is not observed in UHMWPE (Kurtz 2004). Table 2.1 gives the physical and mechanical properties of UHMWPE.

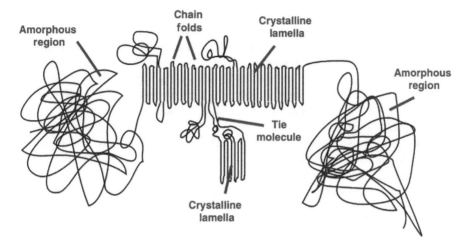

FIGURE 2.2 Schematic of molecular chains in UHMWPE (Kurtz 2004).

TABLE 2.1
Average Values for the Physical and Mechanical Properties of UHMWPE (Kurtz 2004)

Property	Value
Glass transition temperature (°C)	−160
Melting temperature (°C)	125–138
Poisson's ratio	0.46
Density (g/cm³)	0.932–0.945
Elastic modulus (GPa)	0.8–1.6
Yield strength (MPa)	21–28
Ultimate tensile strength (MPa)	39–48
Ultimate tensile elongation (%)	350–525
Crystallinity (%)	39–75

2.2 TRIBOLOGICAL PROPERTIES OF UHMWPE

Tribology is a branch of material science dealing with three main concepts: wear, friction, and lubrication between interacting bodies. These three concepts are in relation to each other. Various factors are affecting the tribological systems such as material properties, surface conditions, contact mechanics, dynamics of a sliding system, environmental conditions, etc. For this reason, the tribological behavior of sliding systems may exhibit unsystematic responses under different conditions, and therefore, the design of tribological systems may come to a state of struggle while optimizing the process parameters. For example, an increase in friction may lead to accelerated wear in materials; however, high wear-resistant materials may exhibit high friction during sliding conditions.

Metal/metal contacts predominate the tribological systems in engineering applications; however, polymer included tribological applications have become prevalent due to the recent developments in polymer science. First, tribological studies on polymers were started with rubbers, which are the main materials in the automotive industry (Roth, Driscoll, and Holt 1942; Schallamach 1952). In 1952, frictional properties of different polymers – polytetrafluoroethylene (PTFE), polyethylene (PE), polystyrene (PS), polymethylmethacrylate (PMMA), and polyvinylchloride (PVC) – were investigated by Shooter (Shooter 1952). After these studies, polymers were recognized as proper materials for tribological applications, and therefore, polymer tribology became more of an issue in the 1960s (Sinha and Briscoe 2009). Emerging of UHMWPE in tribological systems can be traced back to 1962 when John Charnley used this polymer for prosthesis purposes. UHMWPE is still widely used in biomedical applications to take advantage of its low friction and very low clinical wear rates (Kurtz 2004). Figure 2.3 shows an artificial total hip replacement in exploded view. As shown in the figure, the acetabular side is generally produced from UHMWPE (Oonishi et al. 2008).

In addition to medical purposes, UHMWPE is an excellent material for mating surfaces as such in bearings, sliders, seals, and liners (Gürgen 2019; Gürgen, Çelik, and Kuşhan 2019). This polymer is a correct choice in dry sliding conditions such as bushes in toys and instruments. Gears in food, textile, and paper industries are also dry systems due to unacceptable lubrication, and thus, UHMWPE is widely utilized in these sectors. Aircraft control linkage bearings can be another example for dry or rarely lubricated systems, where UHMWPE emerges as a suitable material (Friedrich 2014). Hydraulic, pneumatic, or unlubricated systems also take advantage of this advanced polymer for sliding surfaces. However, UHMWPE usage is restricted by low-temperature applications (Tewari, Sharma, and Vasudevan 1989). Since the melting point of this polymer is quite low, high-temperature applications result in the degradation of molecular structures in UHMWPE.

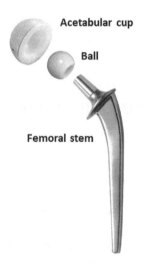

Acetabular cup

Ball

Femoral stem

FIGURE 2.3   Components of a hip prosthesis (Oonishi et al. 2008).

The molecular structure of UHMWPE is a determinant of wear properties for this material. This polymer is composed of crystalline lamellas and amorphous regions as depicted in Figure 2.4a. When the material is subjected to a frictional interaction, molecular chains are stretched in the sliding direction. Because molecular chains in crystalline regions are closely packed, dislocations are initiated in amorphous regions due to loose and entangled chains as illustrated in Figure 2.4b. As suggested in previous works (Galetz and Glatzel 2010), the activation energy for the deformation of amorphous regions is only 2%–10% of that for the deformation of crystalline lamellas. Moreover, deformation in amorphous regions is reversible up to a strain rate of 0.4% due to the elastic recovery of molecular chains. The main deformation mechanisms in this material are stretching and shearing of entangled molecular chains in amorphous regions located between crystalline lamellas (Bowden and Young 1974). However, at excessive loadings, crystalline lamellas are also plastically deformed by being induced at slip planes as shown in Figure 2.4c. Hence, inter-planar distances get larger within the microstructure as shown in Figure 2.4d. During this process, UHMWPE chains remain unbroken; however, very large deformations take place in crystalline structures (J. Huang et al. 2013).

UHMWPE is used for low frictional interactions, and the contact surfaces of this polymer are generally accompanied by metallic components in most applications. Since metallic counterparts are much harder than UHMWPE, the wear mechanism is developed by an abrasive effect such as ploughing. The abrasive effect is seen when hard surfaces slide on soft polymer surfaces. This effect remains scratches, gouges, and wear debris on the sliding surface of polymers. Abrasive wear is mostly associated with asperities of counterpart, which plastically deform the polymer surface.

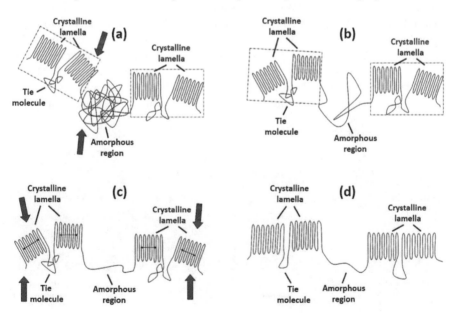

FIGURE 2.4 (a–d) Schematic illustration of molecular structures in UHMWPE during wear deformation (J. Huang et al. 2013).

In ploughing, the contacted surface is pushed ahead and the material flows sideways by forming ridges of a groove. During this process, material removal does not take place significantly; however, there is a displacement of material on the polymer surface. Almost invariably, ploughing is seen together with adhesion, and furthermore, micro-cutting may involve these wear mechanisms during frictional interactions (Sinha and Briscoe 2009). Micro-cutting is observed due to sharp asperities of counterpart, and this mechanism is mostly identical to micro-machining. Material removal is plausible because the polymer surface is cut into small pieces, which produces wear debris like chips. In addition to sharp asperities, the inclusion of hard third-bodies develops the micro-cutting mechanism at the sliding interface. For example, polymer matrix composites may include hard particulate fillers, and filler detachments result in three-body interactions in the tribological systems. Hence, free abrasive particles enhance the wear rate of the polymer surface through micro-cutting. Figure 2.5 shows ploughing and micro-cutting effects on UHMWPE surfaces.

For polymers, adhesion is a much stronger mechanism than abrasive effect because in general, polymers generate transfer films on the contact surface of metallic counterparts (Belyĭ 1982). Transfer films attenuate the direct contact of the counterparts, thereby improving the tribological properties of the sliding system (Qi et al. 2020). At the sliding interface, thin layers of UHMWPE are transferred to a metallic counterpart, and, thus, a thin polymer film is formed on the metallic surface. The driving force in this phenomenon is electrostatic forces, which get stronger between UHMWPE and counterpart in comparison to inside of the bulk UHMWPE. Although transfer films are piled up at sliding interfaces and form wear debris in many polymers, UHMWPE has a unique characteristic in terms of generating stable transfer films. For this reason, UHMWPE in tribological systems provides a stable and low-wear rate sliding interaction. Transfer films of UHMWPE are generally up to 10 μm in thickness while the molecules within the polymer are mostly oriented in the sliding direction. Molecular orientation of sliding direction is also typically observed in bulk UHMWPE. This orientation is responsible for highly low wear rates in UHMWPE at frictional interactions (Wise 1995). Molecular mobility is also a strong factor in the tribological properties because molecular orientation is dependent on mobility.

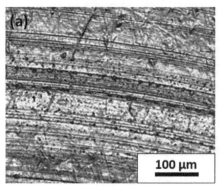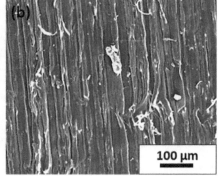

FIGURE 2.5 (a) Ploughing (Gürgen 2019) and (b) micro-cutting (Peng 2013) effects on UHMWPE surfaces.

In addition, perfection and degree of crystallinity have a considerable impact on the tribological behavior of UHMWPE. For example, low-molecular-weight inclusions such as oxygen, oxidation products, or inhibitors are more prone to aggregate in amorphous regions within the microstructure, thereby increasing the heterogeneity in the polymer. For this reason, there is a nonuniform distribution in microstructure, which affects the polymer tribology. Another advantage of transfer film is reduced roughness on counterpart surface. Transferred material from polymer fills the valleys between the asperities of counterpart and thereby developing a smoother surface on the counterpart. Hence, stability of the tribological system is achieved after a certain sliding time of period, and thus, the abrasive effect is suppressed at the contact interface (Sinha and Briscoe 2009). Figure 2.6 shows a metallic counterpart coated by UHMWPE transfer film after a frictional interaction.

Fatigue wear is detected by crack formation, pitting, and delamination by the effect of repeated motions in frictional interactions. Wear debris is observed due to the intersections of cracks growing paths generally perpendicular to sliding directions. Fatigue mechanism develops in long-term frictional interactions, where smooth counterparts slide on polymers without a strong adhesive effect. Deformations in fatigue wear accumulate at surface or subsurface regions contrary to the extension to all body as such in the fatigue of bulk materials. Fatigue-based cracks are originated at the maximum tangential stress points, where friction coefficients exhibit excessive levels. In general, maximum shear stresses are observed at subsurfaces in low friction coefficients ($f < 0.3$) while at high friction coefficients ($f > 0.3$), maximum stress regions shift to the surfaces (Yamaguchi 1990; Totten and Liang 2004). Crack initiation is facilitated by structural defects, which are responsible for stress accumulations within the body. Structural defects can be dents, marks, scratches, and pits on surfaces or inclusions, voids, and cavities in subsurface regions. Under repeated loadings, these defects gradually grow and come close up to each other. Hence, they form larger defects within the structure, thereby leading to material removal from

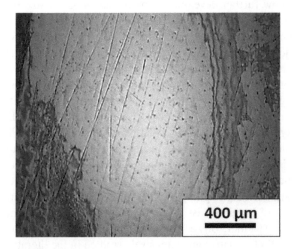

FIGURE 2.6 A metallic counterpart coated by UHMWPE transfer film after a frictional interaction (Trommer et al. 2015).

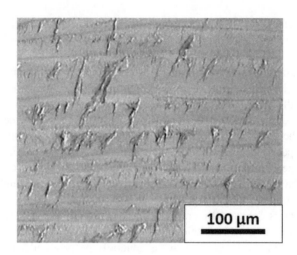

FIGURE 2.7 Fatigue wear cracks on UHMWPE surface after a frictional interaction (Trommer et al. 2015).

the body (Sinha and Briscoe 2009). Figure 2.7 shows fatigue wear cracks on the UHMWPE surface after a frictional interaction.

In recent works, different types of filler materials have been included in the UHMWPE matrix to enhance wear resistance of this polymer. Hard particulate fillers such as carbides, nitrides, and oxides are widely investigated in the UHMWPE matrix due to their superior anti-wear properties. Hard reinforcing materials protect the soft polymer matrix from heavy wear conditions by increasing the material hardness (Gürgen, Sert, and Kuşhan 2021). Particle size comes into prominence in composites having particulate fillers. A well-distributed filler phase is achieved by using finer particles in the polymer matrix, and therefore, mechanical and tribological properties show further development by means of small-size fillers. According to a recent work, nano-size boron carbide particles provide about 15% higher wear resistance than micro-size ones (Sharma, Bijwe, and Panier 2016). However, agglomeration is a major problem in nano-size fillers. To eliminate this drawback, particle loading should be precisely adjusted in the polymer matrix while ensuring homogeneously distributed fillers in the microstructure. In general, a maximum filler amount of 3%–4% is suggested for nano-filler composites (Sharma et al. 2015). Although the adhesive effect predominates over abrasive wear in neat UHMWPE, micro-cutting and micro-ploughing are pronounced in hard particle reinforced UHMWPE composites. In long-term tribological interactions, particle detachment is a common deformation mechanism for UHMWPE composites. Particle removals lead to a third-body interaction in the sliding zone and sometimes accelerate wear formation by means of micro-cutting and micro-ploughing. At this juncture, filler concentration and filler/ matrix adhesion are important issues to lower the particle detachment during frictional interactions (Kumar, Bijwe, and Sharma 2017). In addition to hard particles, fibers in the UHMWPE matrix are extensively studied in the literature. Carbon fiber is one of the options being widely included in the UHMWPE matrix. These fibers are effective to reduce the friction coefficient, thereby contributing to the tribological properties of the polymer matrix (Wood, Maguire, and Zhong 2011). In wet conditions, water absorption of composites slightly increases due to carbon fiber reinforcement, and thus, water lubrication provides development for wear resistance. Fibers also support a certain part of applied load, thereby reducing the contact area between polymer matrix and counterpart (Y. Wang et al. 2017). In addition to carbon fibers, different types of fibers such as glass, basalt, and wollastonite are used as fillers in UHMWPE composites (Tong et al. 2006; Cao et al. 2011; Li et al. 2013).

2.3 A CASE STUDY: WEAR DEFORMATION ON UHMWPE PISTON HEAD IN A GAS SPRING

A gas spring is a simple pneumatic system similar to a mechanical spring in terms of work produced. It uses a compressed gas, generally nitrogen due to its inertness in a cylinder sealed by a piston. When an external loading acts on the piston, the gas is compressed within the cylinder and stores potential energy. After removing the external load, the potential energy is converted into work by extracting the piston from the cylinder. Gas springs are widely utilized in many applications generally to support the weight of a structural member. For example, in the automotive industry, gas springs are incorporated into the struts to support tailgates when they are opened. Machinery, furniture, medical, and aerospace industries also benefit from these simple pneumatic structures in their different applications. Depending on the applications, required force ranges are changed from very light levels to heavy-duty points up to 30 tons in manufacturing and tooling equipment. Figure 2.8 shows the details of a gas spring.

The investigated gas spring in this chapter was removed from a machinery system, which uses a set of gas springs to open flaps on a conveyor band. The failure in the gas spring was detected by a loss in load-bearing capacity as well as extended time period during piston extension. In the inspection of this gas spring, oil leakage was detected around the sealing package. Figure 2.9 shows the detailed views of the failed gas spring. After a preliminary investigation, it was observed that the sealing package was damaged while the piston package was worn due to the sliding motion of the piston during the service life of the gas spring. The O-ring in the piston package provides sealing between the separated rooms by the piston head in the cylinder. The oil acts as a lubricator for the piston package, thereby reducing

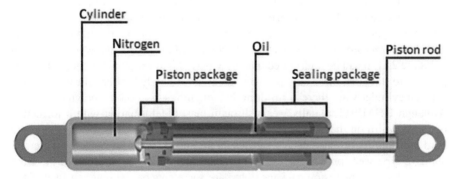

FIGURE 2.8 Components of a gas spring.

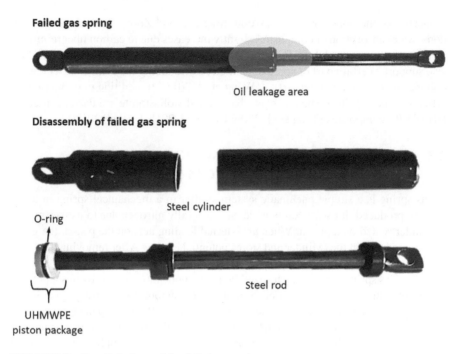

FIGURE 2.9　Detailed views of the failed gas spring.

the friction between the piston package and cylinder walls. However, in the failed gas spring, insufficient lubrication due to the oil leakage resulted in an accelerated wear on the piston package by the effect of increased friction between the mating surfaces. Figure 2.10 shows the microstructural views for the sliding surfaces on the piston package. As shown in the images, there are ploughing tracks on the polymer surface, which are not common in the interface of piston package and cylinder wall. In general, UHMWPE/metal contacts produce friction coefficients of lower than 0.15 (Garino and Willmann 2002), which is a quite low value for the formation of heavy ploughing effect. Upon investigating the residual oil in the gas spring, some contaminations are consisting of solid particles. The contaminants look like sand grains being included inside the cylinder most probably from the external environment during the service life of the gas spring. To identify the contaminant particles, electron dispersive spectroscopy (EDS) analyses were carried out. From the EDS results, the chemical composition of these particles is composed of silicon and oxygen as shown in Figure 2.11. It can be interpreted that these particles are silicon dioxide grains. It is possible to state that these hard particles act as a third body between the piston package and cylinder wall, thereby resulting in ploughing effect on the polymer surface. Although UHMWPE usually wears through plastic deformation with a formation of transfer film on the counterpart while leading to smooth wear tracks, sharp and hard third-bodies produce deep scratches on the soft polymer surface, and therefore, abrasive type wear predominates the frictional interaction. In addition to the wear deformations on the piston package, there are some cuts on the outer surface of the O-ring. It is likely that these formations arise from the particulate inclusions and lead

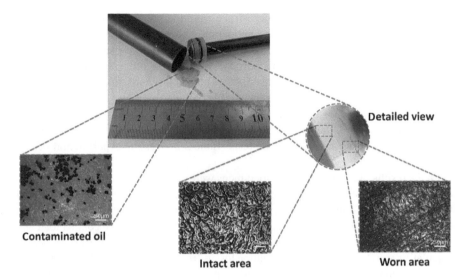

FIGURE 2.10   Microstructural views for the sliding surfaces on the piston package.

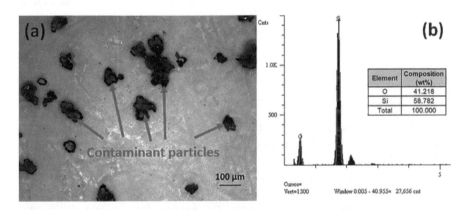

FIGURE 2.11   (a) Contaminant particles and (b) EDS analysis.

to a gradual pressure loss in the compressed nitrogen room of the gas spring. For this reason, this spring shows a reduction in the load-bearing capacity while extending the time period of piston extension.

2.4   WEAR TESTS FOR UHMWPE

In this section, the wear behavior of UHMWPE was investigated in a ball-on-disk configuration. The specimens were produced from UHMWPE powder (43951, Alfa Aesar) having an average particle size of 150 µm according to the manufacturer's specifications. Compression molding method was used to produce the disc-shaped specimens with a diameter of 30 mm and a height of 5 mm. According to early works (S. Wang and Ge 2007; Wu, Buckley, and O'Connor 2002), high molding pressure is suggested to have a consolidated microstructure in the polymer. On the other hand, particular attention is paid to molding temperature because the selected point should be sufficient to fully melt the polymer powder while avoiding thermal degradation due to elevated temperature. For this purpose, the molding process was conducted at a high-pressure level of 350 MPa and 150°C, which is just above the melting point of UHMWPE. A 3 mm diameter tungsten carbide ball was used as a counterpart in the wear tests, in which the sliding distance was selected as 1000 m. To investigate the effect of contact force, the applied load was changed from 2 to 10 N while three different sliding speeds were used in the tests. Table 2.2 gives the experimental design in this study.

In the results, wear tracks on the specimens were analyzed by using optical microscopy. In addition to the visual investigations, specific wear rate (W_s) was calculated for each specimen from Equation 2.1. In this formulation, ΔV is the volume loss from the specimen, L is the contact load, and d is the sliding distance. The contact load and sliding distance are the known inputs in the experimental setup; however, we need a calculation step before finding the volume losses from the specimens. In this step, the cross-sectional areas of wear tracks were multiplied by the length of wear tracks to find the volume losses from the specimens. To do this, the cross-sectional areas were calculated in software, OriginPro9.0, by using the surface profile measurements from the wear tracks with a Mitutoyo SJ-400 surface profilometer. Friction coefficients were also discussed in the results.

$$W_s = \frac{\Delta V}{Ld} \left[\frac{mm^3}{Nm} \right] \tag{2.1}$$

Figure 2.12 shows the cross-sectional areas of the wear tracks on the specimens. As shown in the schematics, there is an obvious increase in the cross-sectional area by increasing the contact force during the wear tests. At the lowest loading condition, which is the case with the specimen PE-2N-20, the maximum Hertzian contact pressure is around 62 MPa. This is a quite high level for UHMWPE, and therefore, the polymer exhibits plastic deformation under this loading. For this reason, the point contact between the components turns to areal contact. Moreover, the diameter of the contact area increases by increasing the applied load. In this light, it is possible to mention that frictional interaction between the contacted surfaces gets larger under

TABLE 2.2
Experimental Design in the Wear Tests

Specimen Code	Contact Load (N)	Sliding Speed (cm/s)
PE-2N-20	2	20
PE-5N-20	5	20
PE-10N-20	10	20
PE-10N-10	10	10
PE-10N-5	10	5

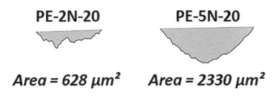

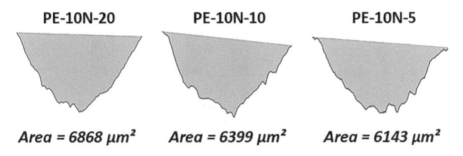

FIGURE 2.12 Cross-sectional areas of the wear tracks on the specimens.

heavy loadings, thereby leading to wider wear tracks on the specimens. On the other hand, the effect of sliding speed is quite lower in comparison with the load effect. Despite the slight deviation, there is a gradual increasing trend in the cross-sectional area by increasing sliding speed. This can be associated with the accelerated adhesive wear by the effect of increasing heat generation as sliding speed increases in the process. Similar results are seen in the specific wear rates as shown in Figure 2.13. Considering the contact load effect, there is a sudden jump in the specific wear rate by applying higher forces. The specific wear rate of the specimen under 5 N is 1.5 times higher than that under 2 N. This increase is around twofold by changing the applied load from 2 to 10 N. Regarding the sliding speed, the specific wear rates are in a narrow bandwidth such as varying from 3.09×10^{-5} to 3.45×10^{-5} mm³/Nm. Although these values are close to each other, there is a systematic variation in the results with respect to sliding speed.

Figure 2.14 shows the optical microscopy images of the wear tracks on the specimens. Wear deformation gets larger by increasing contact load in the tests. The width of the wear track significantly increases under heavy contact loads. As aforementioned in the previous paragraph, plastic deformation on the UHMWPE surface extends to a large scale under heavy loadings and, therefore, contact area increases at the sliding interface. From the micrographs, abrasive effect predominates the wear characteristics of the polymer; however, adhesive type wear is also visible in the wear tracks. The ploughing tracks indicate abrasive wear while the adhesive effect is determined by the exfoliations in the wear tracks and the lamellar material build-ups at the ridges. In the wear tests, the sliding interaction begins with two-body contact, and as the process goes on, frictional interaction results in heat generation in the wear track. Owing to the increased temperature, the specimen shows thermal softening, thereby producing wear debris sticking on the counterpart. Wear debris is involved in the sliding system as a third body. This process leads to material removal areas on

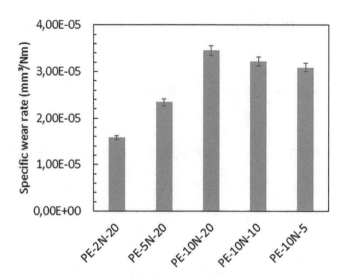

FIGURE 2.13   Specific wear rates for the specimens.

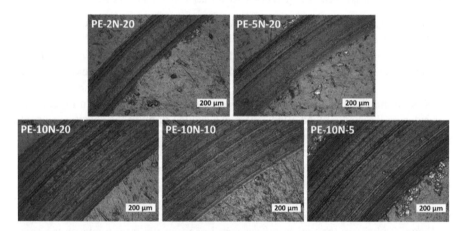

FIGURE 2.14   Micrographs of the wear tracks on the specimens.

the wear track base, which is labeled as exfoliation in Figure 2.15. During the sliding system, the wear debris on the counterpart is fed from the wear track due to the sticky nature of polymer. As the material pile-up reaches an excessive amount on the counterpart, this debris is smeared toward the track ridges as shown as pile-ups in Figure 2.15. This mechanism indicates the adhesive type wear in the sliding system. In addition to the adhesive effect, deep scratches along the wear track due to ploughing indicate that abrasive type wear is also predominant in the frictional interaction as shown in Figure 2.15.

Figure 2.16 shows the mean friction coefficients for the specimens. From this chart, the friction coefficients are varied from 0.049 to 0.082. This shows that each case produces very low friction between the contacted bodies. Considering the cases

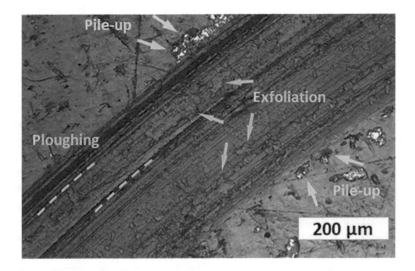

FIGURE 2.15 Detailed view of the wear track on the specimen PE-5N-20.

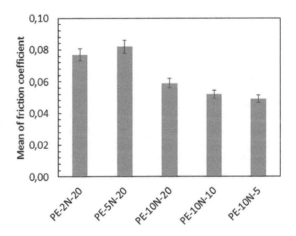

FIGURE 2.16 Mean of friction coefficients for the specimens.

with different contact loads, there is not a systematic variation in the friction coefficients. Although the friction coefficients under the applied loads of 2 and 5 N are quite close to each other, under 10 N, there are reductions of 28% and 23% in the friction coefficient with respect to the cases with 2 and 5 N, respectively. On the other hand, there is a decreasing trend in the friction coefficient by reducing the sliding speed in the wear tests. This variation is attributed to the adhesive effect that at higher speeds, heat generation increases in the wear track. For this reason, thermal softening gets stronger as sliding speed increases in the interaction and consequently accelerating the material removal from the polymer. The removed material acts as wear debris in the wear track, thereby acting as a third body in the system. Hence, frictional interactions are increased between the contacted bodies.

2.5 CONCLUSIONS

This chapter discusses the wear properties of UHMWPE and within this context and incorporates a failure case of the UHMWPE piston head of a gas spring into the discussion. According to the failure analysis of this part, contaminant particles, most probably sand grains result in accelerated wear on the piston head surface. These hard particles act as a third body at the sliding interface of the piston head and cylinder wall, thereby causing a heavy abrasive effect on the polymer surface. In addition, a set of wear experiments is designed to study the effect of different factors on the wear behavior of UHMWPE such as contact load and sliding speed. The contact load effect is investigated by applying three different loads during the sliding interface while sliding speed is changed from 5 to 20 cm/s in the wear tests. By increasing the loading at the sliding interface, there is a drastic jump in the specific wear rate of the polymer. On the other hand, changing the sliding speed within the investigated range does not change the specific wear rate significantly. Despite unsystematic variation in the friction coefficients, the lower contact loads of 2 and 5 N produce higher frictional interaction than the load of 10 N. Regarding sliding speed, there is a systematic reduction in the friction coefficient by increasing speed; however, the results are in a close range.

REFERENCES

Belyĭ, Vladimir Alekseevich, ed. 1982. *Friction and Wear in Polymer-Based Materials.* 1st ed. Oxford; New York: Pergamon Press.

Bowden, Paul, and Robert Young. 1974. "Deformation Mechanisms in Crystalline Polymers." *Journal of Materials Science* 9 (12): 2034–51. doi:10.1007/BF00540553.

Cao, Shoufan, Hongtao Liu, Shirong Ge, and Gaofeng Wu. 2011. "Mechanical and Tribological Behaviors of UHMWPE Composites Filled with Basalt Fibers." *Journal of Reinforced Plastics and Composites* 30 (4): 347–55. doi:10.1177/0731684410394698.

Friedrich, Klaus. 2014. *Friction and Wear of Polymer Composites.* Amsterdam: Elsevier Science. http://qut.eblib.com.au/patron/FullRecord.aspx?p=1152682.

Galetz, Mathias Christian, and Uwe Glatzel. 2010. "Molecular Deformation Mechanisms in UHMWPE during Tribological Loading in Artificial Joints." *Tribology Letters* 38 (1): 1–13. doi:10.1007/s11249-009-9563-y.

Garino, Jonathan P., and Gerd Willmann, eds. 2002. *Bioceramics in Joint Arthroplasty: Proceedings.* Stuttgart; New York: Thieme.

Gürgen, Selim, Abdullah Sert, and Melih Cemal Kuşhan. 2021. "An Investigation on Wear Behavior of UHMWPE/Carbide Composites at Elevated Temperatures." *Journal of Applied Polymer Science* 138 (16): 50245. doi:10.1002/app.50245.

Gürgen, Selim. 2019. "Wear Performance of UHMWPE Based Composites Including Nano-Sized Fumed Silica." *Composites Part B: Engineering* 173 (September): 106967. doi:10.1016/j.compositesb.2019.106967.

Gürgen, Selim, Osman Nuri Çelik, and Melih Cemal Kuşhan. 2019. "Tribological Behavior of UHMWPE Matrix Composites Reinforced with PTFE Particles and Aramid Fibers." *Composites Part B: Engineering* 173 (September): 106949. doi:10.1016/j.compositesb.2019.106949.

Huang, Anmin, Rongjin Su, and Yanqing Liu. 2013. "Effects of a Coupling Agent on the Mechanical and Thermal Properties of Ultrahigh Molecular Weight Polyethylene/Nano Silicon Carbide Composites." *Journal of Applied Polymer Science* 129 (3): 1218–22. doi:10.1002/app.38743.

Huang, Jie, Shuxin Qu, Jing Wang, Dan Yang, Ke Duan, and Jie Weng. 2013. "Reciprocating Sliding Wear Behavior of Alendronate Sodium-Loaded UHMWPE under Different Tribological Conditions." *Materials Science and Engineering: C* 33 (5): 3001–9. doi:10.1016/j.msec.2013.03.030.

Kumar, Anupam, Jayashree Bijwe, and Sanjeev Sharma. 2017. "Hard Metal Nitrides: Role in Enhancing the Abrasive Wear Resistance of UHMWPE." *Wear* 378–379 (May): 35–42. doi:10.1016/j.wear.2017.02.010.

Kurtz, Steven M. 2004. *The UHMWPE Handbook: Ultra-High Molecular Weight Polyethylene in Total Joint Replacement*. Amsterdam; Boston, MA: Academic Press.

Li, Du-Xin, Yi-Lan You, Xin Deng, Wen-Juan Li, and Ying Xie. 2013. "Tribological Properties of Solid Lubricants Filled Glass Fiber Reinforced Polyamide 6 Composites." *Materials & Design* 46 (April): 809–15. doi:10.1016/j.matdes.2012.11.011.

Oonishi, Hiroyuki, Sokchol Kim, Hiroyuki Oonishi Jr., Shingo Masuda, Masayuki Kyomoto, and Masaru Ueno. 2008. "Clinical Applications of Ceramic-Polyethylene Combinations in Joint Replacement." In *Bioceramics and Their Clinical Applications*, pp. 699–717. Elsevier. doi:10.1533/9781845694227.3.699.

Peng, Chang Boon. 2013. "Improving the Wear Resistance of Ultra-High-Molecular-Weight Polyethylene." *SPE Plastics Research Online*, May. doi:10.2417/spepro.004882.

Qi, Huimin, Guitao Li, Ga Zhang, Gen Liu, Jiaxin Yu, and Ligang Zhang. 2020. "Distinct Tribological Behaviors of Polyimide Composites When Rubbing against Various Metals." *Tribology International* 146 (June): 106254. doi:10.1016/j.triboint.2020.106254.

Roth, Frank, Raymond Driscoll, and William Holt. 1943. "Frictional Properties of Rubber." *Rubber Chemistry and Technology* 16 (1): 155–77.

Schallamach, Adolf. 1952. "Abrasion of Rubber by a Needle." *Journal of Polymer Science* 9 (5): 385–404. doi:10.1002/pol.1952.120090501.

Sharma, Sanjeev, Jayashree Bijwe, and Stephane Panier. 2016. "Assessment of Potential of Nano and Micro-Sized Boron Carbide Particles to Enhance the Abrasive Wear Resistance of UHMWPE." *Composites Part B: Engineering* 99 (August): 312–20. doi:10.1016/j.compositesb.2016.06.003.

Sharma, Sanjeev, Jayashree Bijwe, Stephane Panier, and Mohit Sharma. 2015. "Abrasive Wear Performance of SiC-UHMWPE Nano-Composites – Influence of Amount and Size." *Wear* 332–333 (May): 863–71. doi:10.1016/j.wear.2015.01.012.

Shooter, Kenneth. 1952. "Frictional Properties of Plastics." *Proceedings of the Royal Society of London. Series A. Mathematical and Physical Sciences* 212 (1111): 488–91. doi:10.1098/rspa.1952.0250.

Sinha, Sujeet K., and Brian J. Briscoe, eds. 2009. *Polymer Tribology*. London : Singapore ; Hackensack, NJ: Imperial College Press; Distributed by World Scientific.

Tewari, Upendra, Sanjeev Kumar Sharma, and Padma Vasudevan. 1989. "Polymer Tribology." *Journal of Macromolecular Science, Part C: Polymer Reviews* 29 (1): 1–38. doi:10.1080/07366578908055162.

Tong, Jin, Yunhai Ma, R.D. Arnell, and Luquan Ren. 2006. "Free Abrasive Wear Behavior of UHMWPE Composites Filled with Wollastonite Fibers." *Composites Part A: Applied Science and Manufacturing* 37 (1): 38–45. doi:10.1016/j.compositesa.2005.05.023.

Totten, George E., and Hong Liang, eds. 2004. *Mechanical Tribology: Materials, Characterization, and Applications*. New York: Marcel Dekker.

Trommer, Rafael Mello, Marcia Marie Maru, Waldyr Lopes Oliveira Filho, Valder Nykanen, Cristol Gouvea, Bráulio Soares Archanjo, Elano Martins Ferreira, Rui Silva, and Carlos Alberto Achete. 2015. "Multi-Scale Evaluation of Wear in UHMWPE-Metal Hip Implants Tested in a Hip Joint Simulator." *Biotribology* 4 (December): 1–11. doi:10.1016/j.biotri.2015.08.001.

Wang, Shibo, and Shirong Ge. 2007. "The Mechanical Property and Tribological Behavior of UHMWPE: Effect of Molding Pressure." *Wear* 263 (7–12): 949–56. doi:10.1016/j. wear.2006.12.070.

Wang, Yanzhen, Zhongwei Yin, Hulin Li, Gengyuan Gao, and Xiuli Zhang. 2017. "Friction and Wear Characteristics of Ultrahigh Molecular Weight Polyethylene (UHMWPE) Composites Containing Glass Fibers and Carbon Fibers under Dry and Water-Lubricated Conditions." *Wear* 380–381 (June): 42–51. doi:10.1016/j.wear.2017.03.006.

Wise, Donald L., ed. 1995. *Encyclopedic Handbook of Biomaterials and Bioengineering.* New York: Marcel Dekker.

Wood, Weston, Russ G Maguire, and Wei Hong Zhong. 2011. "Improved Wear and Mechanical Properties of UHMWPE–Carbon Nanofiber Composites through an Optimized Paraffin-Assisted Melt-Mixing Process." *Composites Part B: Engineering* 42 (3): 584–91. doi:10.1016/j.compositesb.2010.09.006.

Wu, Jun Jie, Paul Buckley, and John O'Connor. 2002. "Mechanical Integrity of Compression-Moulded Ultra-High Molecular Weight Polyethylene: Effects of Varying Process Conditions." *Biomaterials* 23 (17): 3773–83. doi:10.1016/S0142-9612(02)00117-5.

Yamaguchi, Yukisaburō. 1990. *Tribology of Plastic Materials: Their Characteristics and Applications to Sliding Components.* Tribology Series 16. Amsterdam; New York: Elsevier.

3 Tribology of Spray-Formed Aluminum Alloys and Their Composites

S. K. Chourasiya
MITS

G. Gautam
IIT Roorkee

N. Kumar
BIET

A. Mohan and S. Mohan
IIT (BHU)

CONTENTS

DOI: 10.1201/9781003097082-3

3.1   INTRODUCTION

The aluminum/aluminum alloys are conventional lightweight materials for structural implementations due to their inherent performance such as low density, excellent corrosion resistance in water, petrochemicals, and many chemical systems [1]. In addition, for other applications such as in automobile, aerospace, and marine industry, these properties are not sufficient. They also require the excellent mechanical and tribological performance accompanying the properties mentioned above [2], which is generally exhibited by the composites. Composites are an important category of engineering materials. These materials consist of two or additional different materials that have distinct physical, chemical, and mechanical properties [3]. In composites, one material is a continuous phase; however, other material or materials is a discrete phase. The continuous one is ductile and binds the discrete phase in a particular orientation and position, which transfers the load to the discrete phase. This phase is called the matrix phase, while the discrete phase has high mechanical properties, which improve the properties of the continuous phase. This discrete phase is called the reinforcement phase for the continuous phase in the composites. In general, the properties of composites are superior to the properties of their materials, which have been used to fabricate composite materials [4]. There are different ways to classify composite materials in which the most common way is based upon the type of matrix material. There are metal matrix composites (MMCs), organic matrix composites (OMCs), and ceramic matrix composites (CMCs) [5].

The aluminum matrix composites (AMCs) are the prime engineering composite materials of the MMCs category, which have been very popular in the last few decades because of their lower density, relatively inexpensive, high strength and stiffness, good properties at elevated temperature and damping capacity, high anti-wear life, and good thermal properties [6–7]. These AMCs are generally utilized to manufacture suitable components in different industries such as aerospace, marine, sports and automotive-like in engine parts: cylinder blocks, cylinder liners, pistons, piston insert rings, and brake drums [8–9]. In the AMCs, aluminum and its alloys are used as a continuous matrix material, while generally ceramics (e.g., SiC, B_4C, Al_2O_3, TiC, TiO_2, SiO_2, TiB_2, ZrB_2, Si_3N_4, BN, AlN, and fly ash), intermetallics (like Al_3Ni, Al_3Fe, Al_3Ti, and Al_3Zr) are used as a reinforcement material [10–12]. Besides that, the other reinforcement materials such as solid lubricants, for example, tin (Sn), molybdenum disulfide (MoS_2), tungsten disulfide (WS_2), graphite (C), and lead (Pb) are also used to improve the properties of the AMCs [13]. These reinforcement materials are used in single or hybrid forms in composite materials [14–15]. The AMCs can be classified based on reinforcement morphology. However, the particle-reinforced AMCs (PRAMCs) are generally preferred in the above-mentioned applications because these composites exhibit isotropic properties throughout the volume [16–18].

Among many synthesis techniques to produce aluminum alloys and their composite materials, spray-forming technique is a relatively preferred one because this technique can produce near-net shape material with improved properties and performance [19]. In addition, this technique also exhibits fast metal deposition and solidification rates, which help to reduce the segregation and oxide content and refine the morphology of particles [19]. This technique is also known as the spray casting, spray processing, and spray deposition technique. This technique is a single-step technique that has the sequential stages of atomization of liquid metal in the inert atmosphere, droplet consolidation, and deposition on a substrate or collector [20–22]. Figure 3.1 shows the schematic illustration of the spray-forming technique with distinct parts to fabricate the aluminum alloys and their composites [23]. Different aluminum alloys and their composites are developed through this spray-forming technique with enhanced mechanical and tribological performance [24–28]. These alloys and composites are utilized in a broad range of applications in different industries such as aerospace, automotive, electrical, and electronic packaging. Besides to that, these materials are also used as bearing materials and tribo-constituents for tribological applications [29].

Tribology is explained as the study of three parameters, namely wear, friction and lubrication. It is a fact that due to wear and friction, the components under relative motions like sliding, rolling, etc., in automobile, aerospace, and marine industries encounter huge losses in the form of materials and energy. These losses could be reduced either by the replacement of components by new materials having an

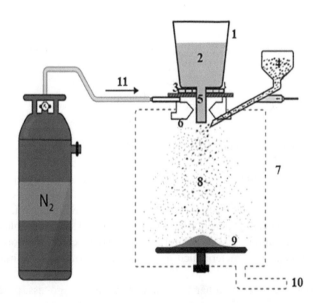

1. Graphite crucible, 2. Molten Al-Si alloy, 3. Electric heater, 4. Graphite powder, 5. Atomizer, 6. Nozzle, 7. Process chamber, 8. Spray cone, 9. Copper substrate, 10. Cyclone, 11. Atomization gas direction

FIGURE 3.1 Schematic illustration of the spray-forming technique with distinct parts to fabricate the aluminum alloys and their composites [23].

outstanding mechanical and tribological performance like lightweight aluminum alloys and their composites or by the application of suitable surface treatment/coating [5]. The tribology of the aluminum alloys and their composites depends on several factors, which include the test operating factors such as sliding distance, applied load, etc., material factors such as mechanical properties, type and amount of reinforcement, physical factors, etc. [30–32]. Therefore, in this chapter, some of the above-mentioned parameters, which influence the wear and friction performance of spray-formed aluminum alloy and their composites are discussed.

3.2   WEAR AND FRICTION BEHAVIOR

The wear is characterized as a continuous loss of material under respective movement into the two bodies. Wear behavior of the spray-formed aluminum alloys and their composites generally displays in terms of cumulative volume loss and wear rate. The value of the wear rate is quantified by the division of volume loss by the sliding distance. However, the volume loss is quantified by the mathematical equation, as shown in Equation 3.1 [33]. The friction is the resistive force of an object that encounters under the motion over another. The friction behavior is defined in terms of friction coefficient or the coefficient of friction. The value of coefficient of friction is calculated by the friction force between bodies and the applied force, which presses together. It can be expressed mathematically in Equation 3.2 [12].

$$\text{Volume loss} = \frac{\text{Mass loss}}{\text{Density}} \times 1000 \qquad (3.1)$$

The volume loss, mass loss, and density are in millimeter cube (mm^3), gram (g), and gram per centimeter cube (g/cm^3), respectively.

$$\text{Coefficient of friction} = \frac{\text{Friction force}}{\text{Applied force}} \qquad (3.2)$$

The coefficient of friction is a number having no physical unit. However, the friction force and the applied force are in newton (N).

Different parameters control the tribology of spray-formed aluminum alloys and their composites. These are [30–31]:

(1) Test operating parameters such as sliding distance, applied normal load, etc. (2) Material parameters like morphology of the matrix phase, morphology (shape and size), type and amount of the reinforcement, and mechanical properties. (3) Physical (environment) parameters such as atmospheric condition (humidity, etc.) and temperature.

Besides to that, the other parameters such as a secondary process, process variables, etc., are also the controlling parameters for the tribology of material through changes in the morphology of the particle and matrix and the mechanical properties [32]. In the following subsections, the influence of some parameters on the wear and friction behavior of spray-formed aluminum alloy and their composites is discussed.

3.2.1 TEST OPERATING PARAMETERS

These parameters of wear and friction test are sliding distance, applied load, etc., which affect the tribological properties of spray-formed aluminum alloys and their composites. These have been explained in detail by responsible mechanisms and worn surface analysis.

3.2.1.1 Effect of Sliding Distance

The volume loss of material with sliding distance generally consists of two different regimes: (1) running-in and (2) steadystate. The steadystate has occurred when the input energy owing to sliding is stabilized with the alteration of mating material surfaces to provide wear condition and it is controlled by the applied load [34]. The increasing tendency of volume loss with sliding distance has been noticed and studied by many material researchers for aluminum alloys and their composites [11,35]. The alteration in volume loss with the sliding distance of spray-formed composite reinforced with graphite particles is shown in Figure 3.2 [36]. It illustrates a linear relationship with the sliding distance. However, the regime regarding running in wear is absent in the plot of volume loss with sliding distance. It may be due to the less number of wear test experiments at a lower sliding distance. The linear alteration of volume loss with sliding distance denotes the steadystate regime.

During sliding under the wear test, the input energy is consumed in two ways: first, plastic deformation followed by removal of material; and second, overcoming the frictional force. The friction force is responsible for the heating of the mating material surfaces. At the initial condition, it is assumed that the temperature of the contact surface is less and wear occurs mainly by the fragmentation of asperities which occurs by the action of cutting and ploughing caused by the penetration of hard asperities (counter steel disc) into the soft asperities or surface (aluminum base materials). This is called the running-in wear, and in this condition, the frictional force and the volume loss of material are high. As the sliding continues, the temperature increases at the interface due to frictional heating, and material gets soft and

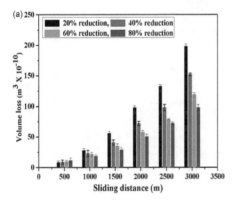
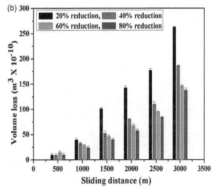

FIGURE 3.2 The volume loss of spray-formed Al-Si alloy composite consisted of graphite particles with varying sliding distance at (a) 20 N and (b) 40 N load. Testing condition; sliding velocity −1.06 m/s [36].

asperities deform easily. When the temperature gets stable through the dissipation of heat through sample, then this condition is called the steadystate and the wear is steadystate wear [37].

Figure 3.3 shows the worn surfaces of spray-formed composite reinforced with graphite particles after different sliding distances. At a small distance (1000 m), the worn surface exhibits shallow grooves and less plastic deformation; but at the large distance (3000 m), it shows the deep grooves, high plastic deformation, and debris particles. It denotes that the wear is mild in condition at a small distance while severe at a large distance.

The coefficient of friction with sliding distance exhibits the steadystate condition after an initial period in the spray-formed aluminum base materials [37]. It has a higher value in the initial period (running-in distance) than the steadystate. Initially, the wear of the material is affected by topographical features (peaks and valleys) of the mating surfaces. These features on the sample create the higher pressure on the peaks and make the cold welding of the mating materials, which give the adhesive part to the coefficient of friction. Hence, this makes the higher coefficient of friction in the initial period than the other.

3.2.1.2   Effect of Normal Load

Normal load is the main test operating variable that significantly influences the wear rate and friction coefficient of aluminum base materials. In general, with increasing normal load, the wear rate increases. The increasing trend of wear rate with normal load has been observed by different material researchers for aluminum base materials [22,38–40]. This type of tendency of wear rate with applied load has also been seen in spray-formed Al-Si alloy composite consisting of graphite particles [36], as shown in Figures 3.2a and b.

Many factors influence the wear rate of spray-formed aluminum base materials. These are broken reinforcement or debris particles, increased real area of contact,

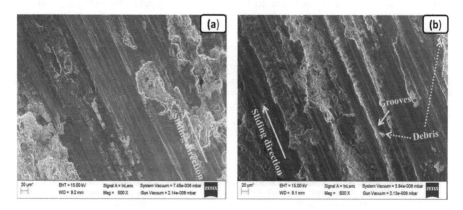

FIGURE 3.3   SEM images revealing the surfaces after wear test of spray-formed Al-Si alloy composite consisted of graphite particles at 20% thickness reduction with varying sliding distance (a) 1000 m and (b) 3000 m. Testing conditions; applied load of 20 N and sliding velocity of 1.06 m/s.

and changed wear mechanism. At high load conditions, the particles in the aluminum matrix may break and increase the wear rate through third body wear [38].

The material surfaces to investigate the wear and friction behavior are not perfectly smooth. They always have some roughness, which indicates that the surface of materials has peaks and valleys. During the wear test, the actual contact occurs at the peaks. That means the real area is different than the apparent area. The real area is the total area of the asperities, which are really in contact. The value of it is always less than the apparent area. Under the application of normal load, the asperities of the surfaces, which are in contact, are plastically deformed and increase the real area, which results in a rise in the wear rate of the aluminum base materials [41].

Apart from this, the temperature at the interface of the mating material surfaces also rises during sliding due to frictional force and an oxide layer generated. At a less load, the generated oxide layer decreases the contact between the material surface and makes an effort to keep the low material loss. Therefore, at a less load, the loss of material is small and the mechanism is mild/oxidative. However, at the high load, the temperature is more as compared to the previous case, which enhances the extent of an oxide layer. This may fracture on continuous sliding and behaves like a third object in the middle of material surfaces and material loss increases. In this condition, the wear mechanism alters from mild/oxidative to severe/oxidative-metallic [32]. The critical value of the normal load whereon the wear mechanism changes from mild/oxidative to severe/oxidative-metallic depends on types of aluminum matrix composites.

Figure 3.4 exhibits the SEM micrograph of the worn surface of spray-formed Al-Si alloy composite consisted of graphite particles at varying loads (20 and 40 N). The SEM micrograph at 20 N load consists of shallow grooves and mild deformation of the surface. However, at the 40 N load, the SEM micrograph shows deep grooves, debris particles, and cracks, which indicates that the wear mechanism changes from mild/oxidative (at 20 N) to severe/oxidative-metallic at high load (40 N). In addition to that, the debris analysis at high normal conditions depicts that the debris particles are large with different morphologies and different colors as shown in Figure 3.5.

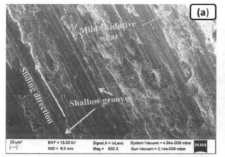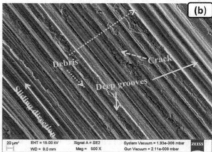

FIGURE 3.4 SEM micrographs of the surfaces after wear test of spray-formed Al-Si alloy composite reinforced with graphite particles at 20% thickness reduction on (a) 20 N and (b) 40 N load. Testing conditions: sliding distance of 3000 m and sliding velocity of 1.06 m/s.

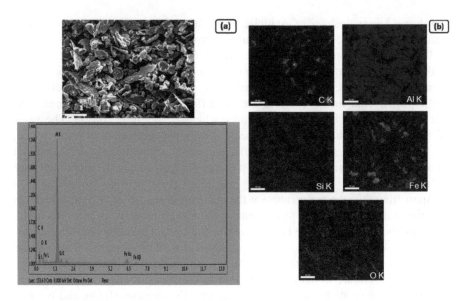

FIGURE 3.5 (a) SEM micrograph with EDS analysis and (b) EDS mapping of debris of spray-formed Al-Si alloy composite reinforced with graphite particles. Testing conditions: thickness reduction 20%, sliding distance 3000 m, applied load 40 N, and sliding velocity 1.06 m/s [32].

The EDS spectrum and elemental mapping of the debris particles display the existence of different elements such as aluminum, silicon, carbon, oxygen, and iron, which validate the severe/oxidative-metallic wear mechanism at high load and the huge amount of material loss [32]. The debris particles are generally released from and below the surface of testing pin material under the wear. Many factors affect the generation of debris particles such as adhesion, distortion, rupture of asperities, ploughing of hard entrapped material, etc. [21].

In general, there is no fixed trend between the coefficient of friction and normal load for aluminum base materials. In some materials, the coefficient of friction decreases with an increase in normal load due to the merging of hard particles in the aluminum matrix [42,43]. While in the other material, the coefficient of friction first decreases and attains minima followed by increases with an increase in normal load. This trend between the coefficient of friction and normal load is due to the generation and breaking of the mechanically mixed layer [44]. In addition, the coefficient of friction also shows an entirely different trend, i.e., fluctuating with a normal load. It is due to the formation and removal of oxides periodically or the happening of different complex reactions during the sliding under the wear test [45,46]. However, in the case of spray-formed aluminum alloys and their composites, the coefficient of friction exhibits a decreasing tendency with increasing normal load [32,38]. This may be owing to the generation of transfer film under the wear test. This generated film is observed to be stable for the broad scale of normal load and long period. However, at a very high load, this generated film may break, and it may increase the coefficient of friction [47].

3.2.1.3 Effect of Sliding Speed

The sliding speed is also a main operating variable that governs the wear rate and friction coefficient of spray-formed aluminum base materials. The wear rate at first decreases with an increase in sliding velocity and gets a minimum value at a particular sliding velocity, and further, it increases with an increase in the sliding velocity in the spray-formed aluminum base materials as reported by material researchers [22,37,48].

In the spray-formed immiscible alloys (aluminum-lead system), at an early stage, the wear rate decreases due to the smearing of lead particles at the interface of materials, which acts as a shield for the matrix surface, leading to decreased further wear. However, after that, the wear rate increases because of the instability of lead film at the interface of materials, which results in subsequent seizure of the material surface [22].

However, in the other aluminum alloy systems (Al-Si alloys and Al-Si-Sn alloy), the wear rate decreases up to a specific value with sliding speed is owing to the competing results of strain rate and temperature developed during wear [49]. Besides that, other possible wear mechanisms are based on worn-out particles [50]. The generation of fine equiaxed particles at low sliding velocities, delamination of compacted equiaxed particles at medium sliding velocities, and delamination of the plastically deformed material at high sliding velocities are the responsible phenomena in the above-mentioned behavior of the wear rate with the sliding velocity.

However, according to another theory, the wear rate initially decreases owing to the evolution of the oxide layer at the sample surface, which minimizes the metallic contact and interaction of asperities. While in the later stage, the wear rate increases after the critical sliding speed due to metallic wear. The high sliding velocity rises the temperature of the sample surface, which softens the material and causes plastic distortion of the matrix outcomes in metallic wear [48].

In certain situations, the wear rate decreases continuously with an increase in sliding speed in the aluminum alloy system [51]. This is owing to the competing results of strain rate and temperature development during the sliding. In addition to that, this can also be caused by the microstructure alteration in the alloy. At low sliding velocities, there is sufficient time for the growth of micro welds. This raises the needed force to shear off the micro welds to continue the relative movement and causes a high wear rate. While at higher sliding velocities, there is shorter time available for the micro welds growth, which decreases the required shear force and causes a less specific wear rate [52]. However, the coefficient of friction is more or less constant with the sliding velocity in the spray-formed aluminum base materials [51,52].

3.2.2 MATERIAL PARAMETERS

3.2.2.1 Effect of Hardness

The wear behavior also depends on the hardness of the spray-formed aluminum base materials. The decrease in wear rate with an increase in hardness has been observed by material researchers in spray-formed aluminum materials [38,53]. The correlation between the wear rate and the hardness of the materials is given by Archard's law, which is explained in the following way:

$$Q = k \times \frac{W}{H} \tag{3.3}$$

where Q is the wear rate defined as volume loss of the material per unit sliding distance, k is the wear coefficient, W is an applied load and H is the hardness of the surface material [54]. It shows that the wear rate is inversely proportional to the hardness of the material.

The high hardness of the material decreases the tendency of ploughing under the wear test. The ploughing of the material occurs when the hard asperities of one material interact with the soft asperities of an other material. The low hardness material having soft asperities on the surface exhibits deep and long grooves with a high value of surface roughness. However, the high hardness material consisting of hard asperities shows the shallow and narrow grooves with less value of surface roughness. In this condition, the worn surface displays a smooth surface [55].

3.2.2.2   Effect of Multiple Reinforcements

The wear behavior of spray-formed aluminum base materials is significantly affected by the addition of multiple reinforcements. The material containing multiple reinforcements shows the lowest wear rate than the separate materials consisting of single reinforcement, as observed by the material researcher in the spray-formed aluminum-copper-aluminum oxide-lead-based composites [56]. This is because of the combined results of the multiple reinforcements in the material. The first reinforcement, lead is the solid lubricant, which smears on top of the material surface and constructs the thin film of low shear strength, reducing the direct contact of a metallic surface, while the other reinforcement (aluminum oxide) is hard, which restricts the flow of the matrix material (in the above system). In addition, the fine reinforcement harder particles in the composite act as a reservoir, which retains the solid lubricating layer on the wear surface through mechanical interlocking and decreases the wear rate of the material [57].

3.2.2.3   Amount of Reinforcement Particles

The wear behavior of the spray-formed aluminum base materials also depends on the number of reinforcement particles. In general, with an increase in the number of reinforcement particles, the wear resistance of the aluminum base materials increases. This is mainly owing to the enhancement of properties like strength parameters and hardness of the material with an increasing amount of reinforcement particles [58,59]. In the spray-formed aluminum base materials, the anti-wear property of the material also improves with the number of particles [28]. This behavior of wear resistance with the number of particles affected by an applied load and may change with an increase in applied load. During sliding wear, the hard reinforcement particles keep out the matrix material from wear, while sometimes, these particles come out from the surface and entrap in between the surfaces. Here, the wear rate increases through the third body wear. The matrix of the material can be protected to a maximum extent through the incorporation of the critical weight of reinforcement particles. In addition, the increasing amount of reinforcement particles reduces the plastic deformation at the subsurface level, which finally decreases the adhesive wear of the materials [60].

3.2.2.4 Effect of Solid Lubricant

The wear behavior of the spray-formed aluminum base materials is strongly influenced by solid lubricant particles. The decrease in wear rate and coefficient of friction by solid lubricant particles in the aluminum base materials has been noticed and considered by many material researchers in their investigation under dry sliding wear test [22,37,48]. During a dry sliding wear test, these solid lubricants smear and form a fine film of low shear strength between the surfaces, which behaves as a lubricant and minimizes the direct contact between the mating surfaces. This assists in easy shear between the sliding surfaces, which leads to less friction and wear. An ultra-fine and homogeneous distribution of these particles due to the spray-forming process ensures preferable smearing to avoid metallic interaction during sliding.

3.2.2.5 Effect of Porosity

The porosity in the spray-formed aluminum base materials greatly affects the wear and friction behavior. The wear rate of the spray-formed aluminum base materials decreases with a decrease in the porosity under dry sliding wear [32,61]. The improvement in wear resistance of the material with a decrease in porosity is owing to the additive results of decreased surface roughness and increased mechanical properties of the material. The high porosity in the material increases the possibilities of crack nucleation and linkup of pores and decreases the load-bearing area, which results in weak material leading to a decrease in strength and hardness. In addition, the high value of surface roughness at the high porosity decreases the real area of contact between the surfaces and results, increasing the contact pressure, which promotes the generation of wear debris through asperity-asperity interaction. This formed wear debris accelerates the third body abrasion leading to a higher wear rate of the material. High porosity also increases the coefficient of friction, which may be because of the increase in asperities interaction during sliding of material surfaces that increases the material loss [62].

3.2.3 PHYSICAL PARAMETERS

3.2.3.1 Effect of Environment

The environment is one more variable, which significantly influences the wear and friction behavior of the spray-formed aluminum base materials. The material researcher has performed the tribological test in different environments such as air and vacuum atmosphere [63]. They have observed that the wear resistance of the spray-formed aluminum base materials is more in a vacuum atmosphere than in the air atmosphere, while the coefficient of friction is lower in a vacuum atmosphere than in the air atmosphere. This depends on the width and the spreading of the formed film of the element (reinforcement-lead) on the surface of the test material. The film width over the worn surface in a vacuum atmosphere is very fine than the film width in the air atmosphere while the spreading of the lead film is maximum. An optimum thickness of the lubricating film is about 1 μm [64]. The lead spreads easily on the surfaces of the complete material and produces uniform friction with less scatter and steady film. This impedes the adhesion of material surfaces and enhances the wear resistance and lowers the coefficient of friction.

3.2.3.2 Effect of Temperature

The temperature is also the crucial variable that influences the wear performance of the spray-formed aluminum base materials. Kaur and coworkers [65] have studied the variation in wear behavior with the temperature of the spray-formed aluminum alloy and its composites. They have observed that the variation of wear behavior with temperature depends on the types of materials. In some materials, the wear rate initially exhibits an increasing tendency with an increase in temperature, and after that, it shows a decreasing tendency with further increase in temperature. However, in other material systems, the wear rate first decreases with an increase in temperature upto a critical value and beyond that, an increasing trend follows with a further increase in temperature. This tendency of wear with temperature is affected by the material properties and wear mechanisms. In general, the thermal properties of the constituent materials of the composite may affect the wear behavior of aluminum-based materials [66]. With the high thermal conductivity of the reinforcement particles in the composite, the highest wear resistance of the composite can be observed.

3.2.4 OTHER PARAMETERS

The wear and friction performance of the spray-formed aluminum alloys and their composites are also affected by other parameters such as secondary processes, process variables, fabrication process, etc., as discussed in the following subsections.

3.2.4.1 Effect of Warm Rolling

Warm rolling is a major factor that affects the wear performance of the spray-formed aluminum base materials. The warm rolling significantly decreases the wear rate of spray-formed aluminum base materials [36], which is displayed in Figure 3.6. In addition, the wear rate continuously decreases with an increase in % thickness reduction. This is attributed to the increase in mechanical properties such as tensile strength, % elongation, and hardness. Also, the decrease in grain size of α-aluminum and % porosity contributes to a decrease in wear rate and the coefficient of friction. Figures 3.7 and 3.8 show the variation in mechanical properties and other properties (% porosity & grain size of primary phase) with % thickness reduction in the composite [32,23]. These clearly indicate that the mechanical properties are enhanced by the rolling, and it consistently improves with an increasing % reduction. However, the other properties, such as % porosity and grain size of the primary phase, exhibit the opposite trend. The reduction of grain size of α-aluminum improves the hardness value, which decreases the ploughing of the material, while the improved % elongation of the material restricts the particles in the matrix and lowers the third body wear. Besides that, the decreased porosity reduces the chances of crack nucleation

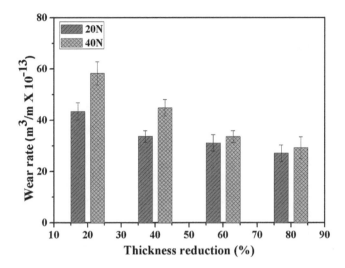

FIGURE 3.6 Effect of thickness reduction (%) on the wear rate of spray-formed Al-Si alloy composite reinforced with graphite particles. Testing conditions: sliding distance of 3000 m and sliding velocity of 1.06 m/s [36].

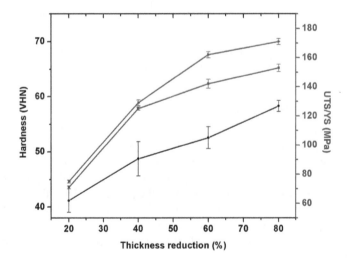

FIGURE 3.7 Hardness, tensile strength, and yield strength of spray-formed Al-Si alloy composite reinforced with graphite particles with a thickness reduction (%) [32].

during sliding while improving the strength of the composite by enhancing the interfacial strength between the particle and the matrix. This also reduces the detachment of the particle, hence, third body wear.

The worn surfaces of the warm rolled aluminum base material after different % thickness reduction are shown in Figure 3.9 [32]. At lower % thickness reduction, the worn surface of the composite exhibits the ruptured oxide layer, deep grooves, high plastic distortion, and fewer particles of solid lubricant in the grooves (Figure 3.9a).

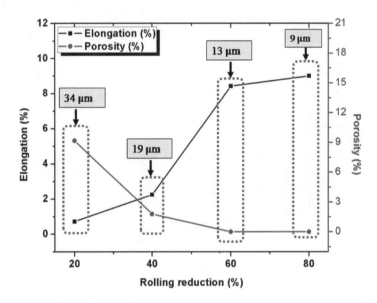

FIGURE 3.8 Variation in elongation (%), porosity (%) with thickness reduction (%) of spray-formed Al-Si alloy composite reinforced with graphite particles. The grain size of the primary phase in composite after different rolling reduction (%) is also shown [23].

Nevertheless, at higher % thickness reduction, the worn surface displays shallow grooves, less plastic distortion, a smooth oxide layer, and more particles of solid lubricant (Figure 3.9d).

3.2.4.2   Effect of Hot Pressing

Hot pressing is also an important factor, which affects the wear performance of the spray-formed aluminum base materials. The hot pressing significantly decreases the wear rate of spray-formed aluminum base materials [67]. This is attributed to reducing porosity and the refinement of the particles and intermetallics presented in the spray-formed composite. The morphology of the particles affects the wear behavior of the aluminum base alloy to a large extent. The coarse particles are broken more frequently compared with fine particles under the sliding, which increases the loss of the material through third body wear. In this condition, the worn surface shows very little deformation.

3.2.4.3   Effect of Surface Treatment

The surface treatment of spay-formed aluminum base materials also affects the wear behavior. Jung and coworkers [68] have studied the wear behavior of the spray-formed aluminum base materials after plasma nitriding and the duplex treatment (plasma nitriding followed by electron beam remelting). They have observed that under dry sliding condition, the wear volume of the spray-formed aluminum base material is significantly reduced after the plasma nitriding and duplex treatment. However, a higher wear resistance is achieved after the duplex treatment. This is due to the destruction of the diffusion-induced cavities beneath the nitride film and

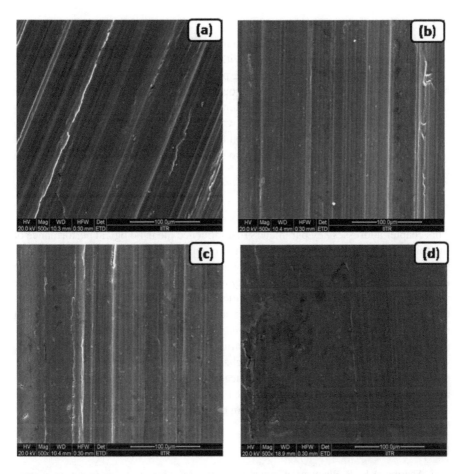

FIGURE 3.9 SEM micrographs of worn surface of spray-deposited Al-Si alloy composite reinforced with graphite particles with thickness reduction (%) (a) 20, (b) 40, (c) 60, and (d) 80. Testing conditions: sliding distance of 3000 m, applied load 20 N, and sliding velocity of 1.06 m/s [32].

enhanced bond strength between the nitride layer and substrate. The increased hardness after the duplex treatment also improves the load-bearing support for nitride film, which also contributes to an improvement in the wear resistance.

3.2.4.4   Effect of Fabrication Process

The wear behavior of the composite is also influenced by the fabrication process. The wear resistance of the composite fabricated by the spray-forming process is significantly higher than the composite fabricated by the stir casting process [38]. This is owing to the drawbacks in microstructural uniformity, porosity, and interfacial bonding in between matrix and the reinforcement particles in the composite fabricated by the stir casting process [69]. In addition, the lower hardness of the composite fabricated by the stir cast process also imparts lower wear resistance than the composite fabricated by the spray-forming process.

3.2.4.5 Effect of Process Variables

The spray process variables are key parameters, which influence the wear performance of the aluminum base materials. With optimization of spray process variables, the high wear resistance material can be produced. Rudrakshi and coworkers [70] have worked on the influence of spray process variable, namely deposition distance on the wear behavior of the Al-Si-Pb system. They have added that at a high deposition distance, the ultrafine lead particles with unvarying size are noticed owing to a reduction in the volume of liquid phase on the substrate. Further, these ultrafine and uniform distributed lead particles improve the process of smearing lead particles on the material surfaces.

3.3 SUMMARY

This chapter addresses key parameters that influence the tribological performance of the spray-formed aluminum alloys and their composites. The operating mechanisms responsible for the alteration in wear and friction with testing parameters like sliding distance, normal load and sliding velocity of different spray-formed aluminum base materials are well discussed with the critical analysis of the worn surface. Besides that, the influence of material parameters such as hardness, porosity, and amount of reinforcement (single and multiple) on wear and friction behavior is also addressed. The other parameters, such as warm rolling, hot pressing, surface treatment, etc., which affect the microstructure and mechanical properties of the materials leading to improved wear and friction performance of spray-formed aluminum base materials are also explained. With proper choice of spray process variables, wear resistance and friction coefficient of the material can be optimized.

REFERENCES

1. Sharma, A.K., R. Bhandari, A. Aherwar, R. Rimašauskienė, C. Pinca-Bretotean, A study of advancement in application opportunities of aluminum metal matrix composites, *Materials Today: Proceedings*, 26 (2), (2020), 2419–2424.
2. R. Gecu, S.H. Atapek, A. Karaaslan, Influence of preform preheating on dry sliding wear behavior of 304 stainless steel reinforced A356 aluminum matrix composite produced by melt infiltration casting, *Tribology International*, 115 (2017), 608–618.
3. K.K. Chawla, *Composite materials, Science and Engineering*, Third edition, Springer, New York, (2013).
4. D.B. Miracle and S.L. Donaldson, *A.S.M Handbook*, Volume 21: Composites, ASM International, Materials Park, OH, (2001).
5. G. Gautam, N. Kumar, A. Mohan, S. Mohan, Tribology of aluminium matrix composites, In: J.P. Davim, ed. *Wear of Composite Materials*, DE Gruyter, Germany, (2018).
6. G. Gautam, A. Mohan, Effect of ZrB_2 particles on the microstructure and mechanical properties of hybrid $(ZrB_2+Al_3Zr)/AA5052$ Insitu composites, *Journal of Alloys and Compounds*, 649 (2015), 174–183.
7. N. Kumar, G. Gautam, R.K. Gautam, A. Mohan, S. Mohan, A study on mechanical properties and strengthening mechanisms of $AA5052/ZrB_2$ insitu composite, *Journal of Engineering Materials and Technology-Transaction of ASME*, 139 (2017), 011002-1-011002-8.

8. G. Gautam, N. Kumar, A. Mohan, R.K. Gautam, S. Mohan, Strengthening mechanisms of $(Al_3Zr_{mp}+ZrB_{2np})$/AA5052 hybrid composites, *Journal of Composite Materials*, 50 (2016), 4123–4133.

9. M. Gautam, G. Gautam, A. Mohan, S. Mohan, Enhancing the performance of aluminium by chromium oxide, *Materials Research Express*, 6 (2019), 126569.

10. N. Kumar, G. Gautam, R.K. Gautam, A. Mohan, S. Mohan, Synthesis and characterization of TiB_2 reinforced aluminium matrix composites: A review, *Journal of the Institution of Engineers (India): Series D*, 97 (2015), 233–253.

11. S. Mohan, G. Gautam, N. Kumar, R.K. Gautam, A. Mohan, A.K. Jaiswal, Dry sliding wear behavior of Al-SiO$_2$ composites, *Composite Interfaces*, 23 (2016), 493–502.

12. G. Gautam, N. Kumar, A. Mohan, R. K. Gautam, S. Mohan, High temperature tensile and tribological behavior of hybrid (ZrB_2+Al_3Zr)/AA5052 insitu composite, *Metallurgical and Materials Transactions A*, 47 A (2016), 4709–4720.

13. H. Tan, S. Wang, Y. Yu, J. Cheng, S. Zhu, Z. Qiao, J. Yang, Friction and wear properties of Al-20Si-5Fe-2Ni-Graphite solid-lubricating composite at elevated temperatures, *Tribology International*, 122 (2018), 228–235.

14. G. Gautam, A.K. Ghose, I. Chakrabarty, Tensile and dry sliding wear behavior of in situ $Al_3Zr+Al_2O_3$-reinforced aluminum metal matrix composites, *Metallurgical and Materials Transactions A*, 46A (2015), 5952–5961.

15. A. Mohan, G. Gautam, N. Kumar, S. Mohan, R.K. Gautam, Synthesis and tribological properties of AA5052 base insitu composites, *Composite Interfaces*, 23 (2016), 503–518.

16. A. Aherwar, A. Patnaik, C.I. Pruncu, Effect of B_4C and waste porcelain ceramic particulate reinforcements on mechanical and tribological characteristics of high strength AA7075 based hybrid composite, *Journal of Materials Research and Technology*, 9 (5), (2020), 9882–9894.

17. N. Kumar, S.K. Singh, G. Gautam, A.K. Padap, A. Mohan, S. Mohan, Synthesis and statistical modelling of dry sliding wear of Al 8011/6 vol. % AlB2 insitu composite, *Materials Research Express*, 4 (2017), 106514.

18. N. Kumar, G. Gautam, A. Mohan, S. Mohan, High temperature tensile and strain hardening behaviour of AA5052/9 vol. % ZrB_2 insitu composite, *Materials Research*, 21 (2018), e20170860.

19. C. Cui, A. Schulz, K. Schimanski, H.W. Zoch, Spray forming of hypereutectic Al–Si alloys, *Journal of Materials Processing Technology*, 209 (2009), 5220–5228.

20. P.S. Grant, Spray forming, *Progress in Materials Science*, 39 (1995), 497–545.

21. P.K. Rohatgi, S. Ray, Y. Liu, Tribological properties of metal matrix-graphite particle composites, *International Materials Reviews*, 37 (1992), 129–152.

22. G.B. Gouthama, S.N. Ojha, Spray forming and wear characteristics of liquid immiscible alloys, *Journal of Materials Processing Technology*, 189 (2007), 224–230.

23. S.K. Chourasiya, G. Gautam, D. Singh, Performance enhancing of spray formed Al/graphite alloy composite by rolling, *Metals and Materials International*, (2019), doi:10.1007/s12540-019-00547-1.

24. S.K. Chourasiya, A. Si, G. Singh, G. Gautam, D. Singh, A novel technique for automatic quantification of porosities in spray formed warm rolled Al-Si-Graphite composite, *Materials Research Express*, 6 (2019), 116565.

25. Y. Zhou, L.I. Jihong, S. Nutt, E.J. Lavernia, Spray forming of ultra-fine SiC particle reinforced 5182 Al-Mg, *Journal of Materials Science*, 35 (2000), 4015–4023.

26. L.A. Bereta, C.F. Ferrarini, C.S. Kiminami, W.J.F. Botta, C. Bolfarini, Microstructure and mechanical properties of spray deposited and extruded/heat treated hypoeutectic Al–Si alloy, *Materials Science and Engineering: A*, 448–451 (2007), 850–853.

27. S. Hariprasad, S.M.L. Sastry, K.L. Jerina, R.J. Lederich, Microstructures and mechanical properties of dispersion-strengthened high-temperature Al-8.5Fe-1.2V-1.7Si alloys produced by atomized melt deposition process, *Metallurgical Transaction A*, 24 (1993), 865–873.

28. F. Wang, H. Liu, Y. Ma, Y. Jin, Effect of Si content on the dry sliding wear properties of spray-deposited Al-Si alloy, *Materials & Design*, 25 (2004), 163–166.

29. K. Raju, S.N. Ojha, A.P. Harsha, Spray forming of aluminium alloys and its composites: An overview, *Journal of Materials Science*, 43 (2008), 2509–2521.

30. V.K. Anand, A. Aherwar, M. Mia, O. Elfakir, L. Wang, Influence of silicon carbide and porcelain on tribological performance of Al6061 based hybrid composites, *Tribology International*, 151 (2020), 106514.

31. A. Mohan, G. Gautam, N. Kumar, S. Mohan, Sustainable materials for tribological applications. In: S. Hashmi, I.A. Choudhury eds., *Encyclopedia of Renewable and Sustainable Materials*, Vol. 1, Elsevier, Oxford, (2020).

32. S.K. Chourasiya, G. Gautam, D. Singh, Mechanical and tribological behavior of warm rolled Al-6Si-3Graphite self lubricating composite synthesized by spray forming process, *Silicon*, 12 (2019), 831–842.

33. G. Gautam, N. Kumar, A. Mohan, S. Mohan, D. Singh, ZrB_2 nanoparticles transmuting tribological properties of $Al_3Zr/AA5052$ composite, *Journal of the Brazilian Society of Mechanical Sciences and Engineering*, 41 (2019), 469.

34. A. Davis, T.S. Eyre, The effect of silicon content and morphology on the wear of aluminium-silicon alloys under dry and lubricated sliding conditions, *Tribology International*, 27 (1994), 171–181.

35. S.A. Kori, T.M. Chandrashekharaiah, Studies on the dry sliding wear behavior of hypo-eutectic and eutectic Al–Si alloys, *Wear*, 263 (2007), 745–755.

36. S.K. Chourasiya, G. Gautam, D. Singh, Influence of rolling on wear and friction behaviour of spray formed Al alloy composites, *Materials Today: Proceedings*, (2020), DOI: 10.1016/j.matpr.2019.12.304.

37. M. Anil, V.C. Srivastava, M.K. Ghosh, S.N. Ojha, Influence of tin content on tribological characteristics of spray formed Al–Si alloys, *Wear*, 268 (2010), 1250–1256.

38. S.K. Chaudhury, A.K. Singh, C.S. Sivaramakrishnan, S.C. Panigrahi, Wear and friction behavior of spray formed and stir cast Al–2Mg–11TiO$_2$ composites, *Wear*, 258 (2005), 759–767.

39. K. Tiwari, G. Gautam, N. Kumar, A. Mohan, S. Mohan, Effect of primary silicon refinement on mechanical and wear properties of a hypereutectic Al-Si alloy, *Silicon*, 10 (2018), 2227–2239.

40. J.P. Pathak, S.N. Ojha, Effect of processing on microstructure and wear charecteristics of an Al-4.5Cu-10Pb alloy, *Bulletin of Materials Science*, 18 (1995), 975–988.

41. P.K. Rohatgi, M.T. Khorshid, E. Omrani, M.R. Lovell, P.L. Menezea, Tribology of metal matrix composites. In: P.L. Menezes et al. eds., *Tribology for Scientist and Engineers: From Basics to Advanced Concepts*, Springer, New York, (2013).

42. G. Gautam, A. Mohan, Wear and friction of AA5052-Al$_3$Zr insitu composites synthesized by direct melt reaction, *Journal of Tribology-Transaction of ASME*, 138 (2015), 021602-1-021602-12.

43. G. Gautam, N. Kumar, A. Mohan, R.K. Gautam, S. Mohan, Tribology and surface topography of tri-aluminide reinforced composites, *Tribology International*, 97 (2016), 49–58.

44. N. Kumar, G. Gautam, R.K. Gautam, A. Mohan, S. Mohan, Wear, friction and profilometer studies of insitu AA5052/ZrB2 composites, *Tribology International*, 97 (2016), 313–326.

45. A. Mandal, B.S. Murty, M. Chakraborty, Sliding wear behavior of T6 treated A356–TiB$_2$ in situ composites, *Wear*, 266 (2009), 865–872.

46. A. Ravikiran, M.K. Surappa, Oscillations in coefficient of friction during dry sliding of A356 Al–30 wt.% SiCp MMC against steel, *Scripta Materialia*, 36 (1997), 95–98.

47. N. Natarajan, S. Vijayaranangan, I. Rajendran, Wear behavior of A356/25SiC$_p$ aluminium matrix composites sliding against automobile friction materials, *Wear*, 261 (2006), 812–822.

48. K.V. Ojha, A. Tomar, D. Singh, G.C. Kaushal, Shape, microstructure and wear of spray formed hypoeutectic Al–Si alloys, *Materials Science and Engineering A*, 487 (2008), 591–596.

49. K.M. Jasim, Nature of subsurface damage in Al–22 wt.% Si alloys sliding dry on steel discs at high sliding speeds, *Wear*, 98 (1984), 183–197.

50. C. Subramanian, Effects of sliding speed on the unlubricated wear behaviour of Al-12.3 wt.%Si alloy, *Wear*, 151 (1991), 97–110.

51. D.M. Goudar, V.C. Srivastava, G.B. Rudrakshi, K. Raju, S.N. Ojha, Effect of tin on the wear properties of spray formed Al–17Si alloy, *Transaction of the Indian Institute of Metals*, 68 (2015), S3–S7.

52. K. Raju, A.P. Harsha, S.N. Ojha, Evolution of microstructure and its effect on wear and mechanical properties of spray cast Al–12Si alloy, *Materials Science and Engineering A*, 528 (2011), 7723–7728.

53. H. Ye, An overview of the development of Al-Si-alloy based material for engine applications, *Journal of Materials Engineering and Performance*, 12 (2003), 288–297.

54. J.F. Archard, Contact and rubbing of flat surface. *Journal of Applied Physics*, 24 (1953), 981–988.

55. S. Kumar, M. Chakraborty, V.S. Sarma, B.S. Murty, Tensile and wear behavior of in situ Al–7Si/TiB$_2$ particulate composites, *Wear*, 265 (2008), 134–142.

56. M. Anil, S.N. Ojha, Spray processing and wear characteristics of Al-Cu-Al$_2$O$_3$-Pb based composites, *Journal of Materials Science*, 41 (2006), 1073–1080.

57. H. Fallahdoost, A. Nouri, A. Azimi, Dual function of TiC nanoparticles on tribological performance of Al/graphite composites, *Journal of Physics and Chemistry of Solids*, 93 (2016), 137–144.

58. N. Kumar, G. Gautam, R.K. Gautam, A. Mohan, S. Mohan, High temperature tribology of AA5052/ZrB$_2$ PAMCs, *Journal of Tribology-Transaction of ASME*, 139 (2017), 011601-1-011601-12.

59. G. Gautam, N. Kumar, A. Mohan, R.K. Gautam, S. Mohan, Synthesis and characterization of tri-aluminide insitu composites, *Journal of Material Science*, 51 (2016), 8055–8074.

60. A. Vencl, A. Rac, I. Bobic, Tribological behavior of Al-based MMCs and their application in automotive industry, *Tribology Industry*, 26 (2004), 31–38.

61. D. Singh, K. Puri, V. Kumar, The influence of rolling process on the porosity and wear behavior of spray-formed Al-6%Si-20%Pb, *IOSR Journal of Mechanical and Civil Engineering*, 10 (2013), 12–17.

62. A. Sinha, Z. Farhat, Effect of surface porosity on tribological properties of sintered pure Al and Al 6061, *Materials Sciences and Applications*, 6 (2015), 549–566.

63. G.B. Rudrakshi, V.C. Srivastav, S.N. Ojha, Microstructural development in spray formed Al-3.5Cu-10Si-20Pb alloy and its comparative wear behaviour in different environmental conditions, *Materials Science and Engineering A*, 457 (2007), 100–108.

64. E.W. Roberts, Thin solid lubricant film in space, *Tribology International*, 23 (1990), 95–103.

65. K. Kaur, O.P. Pandey, High temperature sliding wear of spray-formed solid-lubricated aluminum matrix composites, *Journal of Materials Engineering and Performance*, 22 (2013), 3101–3110.

66. P. Vissutipitukul, T. Aizawa, Wear of plasma-nitrided aluminum alloys, *Wear*, 259 (2005), 482–489.

67. D.M. Goudar, K. Raju, V.C. Srivastava, G.B. Rudrakshi, Effect of copper and iron on the wear properties of spray formed Al–28Si alloy, *Materials and Design*, 51 (2013), 383–390.

68. A. Jung, A. Buchwalder, E. Hegelmann, P. Hengst, R. Zenker, Surface engineering of spray-formed aluminium-silicon alloys by plasma nitriding and subsequent electron beam remelting, *Surface & Coatings Technology*, 335 (2018), 166–172.

69. S.K. Chaudhury, Processing and characterization of Al-TiO2 composites, Ph.D. Thesis, Department of Metallurgical and Materials Engineering, I.I.T-Kharagpur, India, (2001).

70. G.B. Rudrakshi, V.C. Srivastava, J.P. Pathak, S.N. Ojha, Spray forming of Al-Si-Pb alloys and their wear characteristics, *Materials Science and Engineering A*, 383 (2004), 30–38.

4 Ranking Analysis and Parametric Optimization of ZA27-SiC-Gr Alloy Composites Based on Mechanical and Sliding Wear Performance

Ashiwani Kumar
Feroze Gandhi Institute of Engineering and Technology

Mukesh Kumar
Malaviya National Institute of Technology

CONTENTS

DOI: 10.1201/9781003097082-4

4.1 INTRODUCTION

The Al alloy, like ZA27, has a better mechanical frame of the structure, and wear properties were utilized for preparing and replacing traditional materials/selections, e.g., surfaces of sliding like a disk or bearing material. Nowadays, using metal matrix composite (mmcs) materials in automobile industries is growing day by day, leading to enhancing the demand for developing fabricating methods, which can enhance the mechanical and sliding performance of such newly developed composites [1]. "This article scopes to investigate the optimal weight percentage filler (SiC/Gr) into ZA27 for mechanical and dry sliding wear performance meant for bearing applications. Several authors have reported similar results like Liu et al. [2] examined the mechanical and wear performance behavior of aluminum (Al)–magnesium (Mg)–graphite (Gr) composite and seen that the mechanical properties improve with increment in filler content. Karakulak et al. [3] observed an increase in mechanical performance with nickel in aluminum alloy composite. Kumar et al. [4] investigated diminish in the rate of wear with fly ash and graphite-reinforced alloy composite. Karthick et al. [5] observed the maximum rate of wear and mechanical performance of composite enhances with enhanced Al_2O_3/SiC reinforcement. Singh et al. [6] analyzed the performance of coefficient of friction (COF) and wear rate of composite reduction with decrements in SiC filled ZA27 alloy composite. Girisha et al. [7] examined the impact of graphite-filled ZA27 aluminum alloy composites prepared through the stir casting method. It found a decrease in the wear rate of Gr mixed ZA27 alloy composite with increased filler content and improved wear resistance at 6 wt% Gr reinforced ZA27 aluminum alloy composite. Baradeswaran et al. [8] found the mechanical properties (like hardness, tensile, flexural, and compressive) and resistance of wear increase with increasing the filler content. Kumar and Kumar [9] found the mechanical and dry sliding performance improves with an increase in the B_4C/rice rush and filler content for armor application. Goudarzi and Akhlaghi [10] investigated the performance of COF and the rate of wear improved with increment in load. Toptan et al. [11] found the influence of B_4C on different parameters (like velocity of sliding, load, and a distance of sliding) on sliding performance of composite, and worn mechanism of specimens is computed by SEM. Similar results are reported by Ashiwani Kumar and Mukesh Kumar [12], and they found the higher performance of wear rate of composite showed impendence of graphite-reinforced alloy composite. Kumar et al. [13] observed the performance of the tribological and mechanical properties of Ni powder filled with AA7075 composite for gear application. Bhaskar et al. [14] found the material performance of reinforcement (silicon carbide) filled AA-2024 alloy composite is computed through MCDM methods (like AHP and TOPSIS). The reported outcomes are in tune with the raking performance of fabricated composite and prove that the MCDM techniques are used for adept decision in order for the newly developed materials of performance based on evaluation criteria. Kranthi et al. [15] recommended the neural network method through developed the mathematical model to run experiment outcomes with the planning of parametric design and investigated the effective or systematic in expecting the wear response to different experimental conditions."

The novelty of the research work presented here comprises (1) designing, developing, and fabricating hybrid ZA27-SiCGr alloy composites with a particular combination of ceramic reinforcements through high vacuum casting method; (2) exploring the interactive combined effect of both the ceramics on physical, mechanical, and tribological behavior of alloy composites as per ASTM standards; (3) adopting the Taguchi design of experiment for designing sliding wear experimental simulations as well as input control factor (such as sliding distance, sliding velocity, normal load, composition, and environmental temperature) optimizing using ANOVA and ranking of new developed ZSG alloy composite using AHP-TOPSIS method; and (iv) analysis of worn-out surface using SEM/EDXS.

4.2 MATERIALS AND METHODOLOGY

4.2.1 MATERIAL, DESIGN ASPECTS, AND FABRICATION OF ZA27-SiC-GR ALLOY COMPOSITES

Details of ingredients, design aspects, and fabrication procedure of the alloy composites are described in the flow chart as shown in Figure 4.1. The surface morphology and EDX spectrum of the ingredients are shown in Figure 4.2.

4.2.2 PHYSICAL AND MECHANICAL CHARACTERIZATION

The actual/experimental density of design composite specimens was computed through the water displacement method. The experimental density of fabricated specimens was computed as per ASTMD792 standard using the Archimedes principle; the density of theoretical was measured through the rule of mixture method through Equation 4.1 [16].

$$\rho_c = \frac{1}{\left(\dfrac{W_p}{\rho_p}\right) + \left(\dfrac{W_m}{\rho_m}\right)} \tag{4.1}$$

whereas it is expressed that the weight fraction$=w$ and density of friction$=\rho$, respectively.

The density of voids of the fabricated composite sample was computed through utilizing Equation 4.2 [17].

$$\text{Voids fraction} = \frac{\text{Theoretical}\left(\rho_t\right) - \text{Experimental}\left(\rho_e\right)}{\text{Theoretical}\left(\rho_t\right)} \tag{4.2}$$

The Vickersmicro hardness of fabricated composite was computed through the hardness equipment like Walter Uhltesting tester as ASTM E-92 standard [13] with each five test trials are performed on each sample and the results are recorded in the file. For Vickersmicro hardness, tests are selected parameters (like normal

---

Materials:

a) Base material ZA27; supplied by Bharat Aerospace Metals, Mumbai;

b) Solid lubricant Graphite (Gr); supplied by Savita Scientific Prt. Ltd., Jaipur; 98 μm size; 0-6 wt.%;

c) Silicon Carbide (SiC); supplied by Savita Scientific Pvt. Ltd., Jaipur; 10 μm size; 0-6 wt.%;

Design aspects:

ZSG-0 = 0 wt.% SiC and Gr; rest ZA27

ZSG-2 = 2 wt.% SiC and Gr; rest ZA27

ZSG-4 = 4 wt.% SiC and Gr; rest ZA27

ZSG-6 = 6 wt.% SiC and Gr; rest ZA27

Fabrication procedure:

1. The computed quantity of ZA27 alloy rods was cleaned and its small parts were melted in a graphite crucible using a high vacuum induction furnace. The melted material was held at 850°C for 15 minutes; thereafter the value of temperature lowered to 650°C (zone of mushy i.e. between solidus and liquidus temperatures of the base matrix).

2. The metal powder reinforcements were preheated separately at 800°C for 2 hours.

3. In order to improve the wettability of filler phases in the molten melt 2 wt.% Mg powder was added.

4. The achieve uniform mix fully automatic stirrer (stainless steel; speed ~ 300 rpm; time 15 min.) was used.

5. The blend was poured into fixed cast iron mould with dimensions 150 × 90 × 10 mm^3 and permitted to solidify to room temperature in air (normalizing process).

6. The samples may be cut out using wire Electric Discharge Machine(EDM) as per dimensions followed by polishing with emery papers.

FIGURE 4.1   Materials, design aspects, and fabrication method of ZSG alloy composites.

load = 190 kg; time duration = 15 seconds) and performed on specimens. We get from Equation 4.3

$$HV = 1.854 \frac{L}{D} \tag{4.3}$$

Here, it is expressed that L = applied load (kg) and D = diagonal diameter (mm).

The tensile experiment of fabricated composite specimen was conducted on the UTM tester (ASTMD-3039-76)[38] and taken parameter (like the flat shape specimen dimension = 160 × 10 × 10 mm^3; length of span = 75 mm; cross-head speed = 2 mm/s).

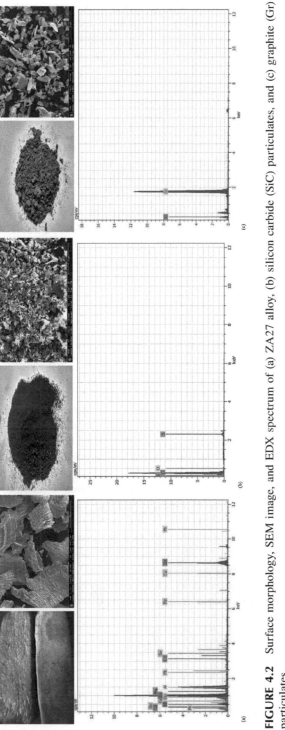

FIGURE 4.2 Surface morphology, SEM image, and EDX spectrum of (a) ZA27 alloy, (b) silicon carbide (SiC) particulates, and (c) graphite (Gr) particulates.

Flexural tests of designing ZSG alloy composite specimens were conducted on a UTM machine parameter (taken as specimen dimensions $= 50 \times 10 \times 10$ mm^3; span length $= 40$ (mm); cross-head speed $= 2$ m/s) followed as per (ASTM D 2344-84) [8] standard. Flexural strength (FS) was computed through using Equation 4.4 [13].

$$F = \frac{3PL}{2bt^2} \qquad (4.4)$$

whereas it is expressed that $P =$ load (kg), $b =$ width of specimen (mm), $t =$ sample thickness of the sample (mm), and $L =$ span length of the sample (mm).

The impact strength (in terms of impact energy) of fabricated composite specimens was computed using an impact tester (ASTM E-23) [13] and taken machine parameter (such as specimen size $= 64 \times 12.8 \times 3.2$ mm^3; depth of notch $= 10.2$ mm). The compressive strength of fabricated composites was computed through using the compression test (ASTM E9-09) [18] and taken parameter (specimen size $= 10 \times 10 \times 10$ mm^3; span length $= 50$ mm; speed $= 2$ m/s), and compression test was performed on UTM.

4.2.3 Dry Sliding Wear Tribometer

The friction and sliding wear performance was measured through using MST (Multi-specimen Tribo-tester: Model: TR-705; Ducom; ASTM G-99) [4] equipment (Figure 4.3) SiC-Gr filled ZA27 alloy composites. The machine parameters (such as rotating disk $= 62$ HRC, EN-31, hardened steel; specimen dimension $= 14 \times 9 \times 10$ mm^3, and load applied to the fixed specimen in the vertical direction) have taken for the experimentations. The steady-state tests were taken parameter (like constant load 20 N; the sliding velocity of 1, 1.25, 1.5, and 1.75 m/s). Sliding wear performance for sliding distance of 700 m and constant disk track diameter of 40 mm was conducted on Tribo-tester and measured the wear data. All the experiments were performed at the environmental conditions and were completed in given duration of time. After each test, the pin and disk surfaces were cleaned with alcohol to eliminate the collected wear debris particles and material moved to sample.

Finally, for each experimental data, the pin type sample was cleaned and previously weighed by an Electronic Balance machine (EBM) (taken $\pm 1 \times 10^{-3}$ mg accuracy). The experimental collection data on wear rate and COF was obtained from the machine, which analyzed data. Therefore, the experimental specific wear rate data are analyzed. The newly developed alloy composite of wear rates (W_s) was obtained by Equation 4.5 [19]

$$W_s = \frac{\Delta m}{\rho \times V_s \times t \times fn} \qquad (4.5)$$

here, it indicates $\Delta m =$ mass loss (g), density (ρ), sliding velocity (V_s), test duration (t), and normal load (fn).

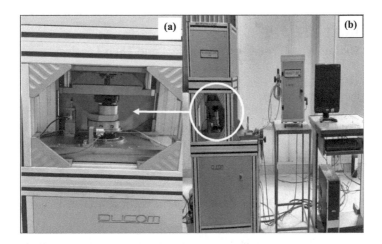

FIGURE 4.3   (a and b) Multi-specimen tribometer.

4.2.4   Taguchi Design of Experiment Optimization

The design structure of the Taguchi exploratory is a technique to reduce the number of preliminaries/cycle and productively accomplish the best outcome for a specific combination. Taguchi methods are developed broadly in the area of engineering and optimization of execution attributes with the permutation of design factors. Therefore, this procedure is associated with the trial and logical idea to compute the factor of the most grounded impact on the outcomes and response for a noteworthy improvement in the entire execution. The attributes of execution in terms of the *S/N* ratio for smallest the better (SB) characteristic are utilized for computing the specific wear rate of SiC/Gr powder-filled ZA27 alloy composite. This article investigated that four parameters (normal load, reinforcement, sliding velocity, and sliding distance) are indicated in Table 4.1, and the design of the L16 orthogonal array is utilized to acquire the ideal parameter combinations. The better characteristics of the *S/N* ratio are indicated in Equation 4.6.

$$\frac{S}{N} = -10\log\frac{1}{Y}\sum X^2 \qquad (4.6)$$

whereas Y=no. of trials/observation and X=computes response of wear rate. Subsequently, the ANOVA strategy was utilized to acquire the noteworthy parameter and ideal execution setting of working parameters [13].

4.2.5   Surface Morphology Studies

The wear micrograph analysis of the worn surface was investigated by utilizing a micrograph and acquired through FESEM equipment. The microstructure behavior of ZA27-SiC-Gr alloy composites was investigated under the below mechanism through SEM.

TABLE 4.1
Sliding Wear Processes Parameters and Their Levels

	Levels			
Input Control Factors	I	II	III	IV
A: Normal load (N)	10	20	30	40
B: Reinforcement content (wt%)	0	2	4	6
C: Sliding velocity (m/s)	1	1.25	1.5	1.75
D: Sliding distance (m)	700	1400	2100	2800

4.2.6 HYBRID AHP-TOPSIS RANKING OPTIMIZATION METHOD

The AHP-TOPSIS method is an efficient multi-criteria decision-making tool that aids scholars in taking the right decision under complex decision situations and is frequently used in areas such as military, economics, inventory management, and material selection. AHP is an adaptable planning and process developed by Thomas Saaty in the 1970s [14] that gives an approach of decision making including the relative performance and order criteria. It includes both conventional and quantitative selection of decisions that reinforced its flexible pertinence. The basic steps are the construction of hierarchy; required priority test; and the confirmation of consistency with the assurance of relative weights[14,20]. The algorithm of the analytic hierarchy process (AHP) is as follows:

Step 1: Construction of hierarchy structure: "The construction of hierarchy structure for intricate decision-making problem enables to comprehend the problem. The deciding target should be in the upper position; criteria of evaluating at the center section and choice (alternatives) lie in the down position of construction of hierarchical structure" as shown in Figure 4.4.

Step 2: Construction of a pairwise comparison matrix: this matrix is prepared by entrusting/passing targets, which depend on human decision, and according to Sattay's 1–9 scale comparison of Pairwise matrix (c), it would appear as:

$$C_{n\times n} = \begin{matrix} & \begin{matrix} C_1 & C_2 & \cdots & C_n \end{matrix} \\ \begin{matrix} C_1 \\ C_2 \\ \vdots \\ C_n \end{matrix} & \begin{bmatrix} 1 & C_{12} & \cdots & C_{1n} \\ C_{21} & 1 & \cdots & C_{2n} \\ \vdots & \vdots & \ddots & \vdots \\ C_{n1} & C_{n2} & \cdots & 1 \end{bmatrix} \end{matrix}$$

here, C_{ij} would be that the degree of choice of (ith criteria (row); jth criteria (column); (i, j value equal to 1, 2, 3, ... , n; no. of the criteria value indicates n). So, the matrix (C) indicates that the square matrix of nth order having target values equals to one diagonal. If it is n-criterion, then it is pairwise comparisons of $\dfrac{n(n-1)}{2}$ such that ($C_{ij} = \dfrac{1}{C_{ji}}$). "If the pair wise comparison matrix (C) = $\left[C_{ij} \right]_{n\times n}$ justified $C_{ij} = C_{ik} \times C_{kj}$ for putting any value ($i, j, k \in 1...n$) then, matrix (C) is said to be consistent; otherwise it

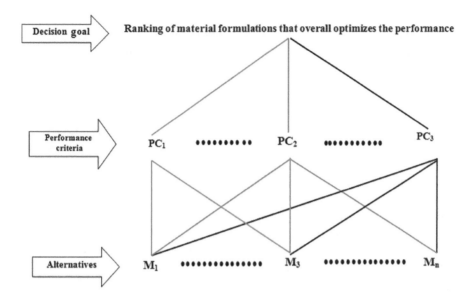

FIGURE 4.4 The hierarchy chart.

indicates inconsistent. Therefore, for a given C of the comparison pairwise matrix (C); the vector of weight (w) may be computed through the CE (characteristic equation): $C \cdot w = \lambda_{max} w$, i.e., w is a vector of the weight of the real total loads or eigenvectors related to the aging esteem and (λ_{max}) indicates the best Eigenvalue estimations from the matrix (C). It demonstrated that for immaculate consistency $(\lambda_{max}) = n$ or rank $= 1$. Quite, an irregularity in needs task may prompt various estimations of λ_{max}; in such cases $\lambda_{max} \sim n$ for more prominent consistency of outcomes. Such type of technique for calculating the vector of the weight of the matrix (C) is alluded to as the chief right eigenvector strategy (saaty; 1980). It is well revealed that a comparison matrix (C) significantly influences the outcomes; subsequently, the test of consistency is performed by determining the index of consistency $(CI) = \dfrac{(\lambda_{max} - n)}{(n-1)}$. (The whole consistency $(\lambda_{max} \cong n$ or rank $= 1$ or $CI = 0))$.

Step 3: Finally, admissible consistency of matrix (C) or the degree of consistency check is assessed by determining consistency proportion $(CR) = \dfrac{CI \, (\text{Consistency Index})}{RI \, (\text{Random Index})}$.

"For steady pairwise comparison matrix (CR) value is greater than equal to 0.1 or 10 percentages; then comparison matrix (C) requires to be remade to decrease the irregularity. Therefore, the calculating (CR) value assesses the consistency of judgment inventor also the consistency of the hierarchical network. Table 4.5 illustrates the irregularity index (RI) values in the matrix (C) with the sequence from 1 to 10. The average of the consistency index of 500 arbitrarily produced matrices through RI" [14].

"The TOPSIS idea is reported by Hwang and Yoon [21–23]; its dominance in calculating genuine life crucial issues of different branches is well defended through previous literature.

The basis step calculations of TOPSIS are followed:

Step 1: Prepare decision matrix: the multi-alternative/choices (e.g., m alternatives/choices) and multi-criteria (say n-criteria) of the issue are indicated in the matrix form (say matrix D of $m \times n$ order).

$$D_{m \times n} = \begin{array}{c} \\ A_1 \\ A_2 \\ \vdots \\ A_m \end{array} \begin{array}{cccc} C_1 & C_2 & \cdots & C_n \\ \left[\begin{array}{cccc} p_{11} & p_{12} & \cdots & p_{1n} \\ p_{21} & p_{22} & \cdots & p_{2n} \\ \vdots & \vdots & \ddots & \vdots \\ p_{m1} & p_{m2} & \cdots & p_{mn} \end{array} \right] \end{array}$$

where $C_1, C_2, \ldots, C_n$ are the n-criteria and $A_1, A_2, \ldots, A_m$ are the m-alternatives

The component p_{ij} indicates the performance value of the id alternative/choices (A_i) with regards to the jth attribute (C_j) here $i = 1, 2 \ldots m$ and $j = 1, 2 \ldots n$.

Step 2: To obtain the criteria in unit less form and to ease between attribute comparison; the passages of the upper matrix are normalized (utilizing Equation 4.7). Hence, obtained standardized/normalized matrix is $R = \{r_{ij}\}$ (of the order $m \times n$).

$$r_{ij} = \frac{p_{ij}}{\left[\sum_{i=1}^{m} p_{ij}^2 \right]^{\frac{1}{2}}}, \text{ here } j = 1, 2, \ldots n \tag{4.7}$$

Step 3: A normalized decision matrix (R) is changed into the V (weighted, normalized decision matrix); $V = \{V_{ij}\}$ (utilizing Equation 4.8).

$$V_{ij} = w_j \times r_{ij} \text{ where, } i = 1, 2, \ldots m; \ j = 1, 2, \ldots, n; \ wj \geq 0; \ \sum_{j=1}^{n} w_j = 1 \tag{4.8}$$

where w_j indicates the relative weights or relative importance of jth criteria. It is calculated by using the AHP method as described in the earlier section.

Step 4: Calculate the positive ideal solution (A^+) and the negative ideal solution (A^-) (following under criteria) that depend on the normalized weight matrix as achieved by Equation 4.9.

$$A^+ = \left(v_1^+, v_2^+, \ldots, v_J^+ \right)$$
$$A^- = \left(v_1^-, v_2^-, \ldots, v_J^- \right) \tag{4.9}$$

whereas

$$v_j^+ = \begin{cases} \max V_{ij}, \text{ if } j \text{ is a benefit criteria or larger-the-better} \\ \min V_{ij}, \text{ if } j \text{ is a cost criteria or smaller-the-better} \end{cases} \text{ whereas } j = 1, 2 \ldots n$$

$$v_j^- = \begin{cases} \max V_{ij}, \text{ if } j \text{ is a benefit criteria or larger-the-better} \\ \min V_{ij}, \text{ if } j \text{ is a cost criteria or smaller-the-better} \end{cases}$$

Step 5: To determine the distance of Euclidian (D) between every alternative and the positive ideal solution or the negative ideal solution by utilizing Equation 4.10:

$$D_i^+ = \sqrt{\sum_{j=1}^{n} \left(v_j^+ - v_{ij} \right)^2} \quad \text{where } i = 1, 2, \ldots, m$$

$$D_i^- = \sqrt{\sum_{j=1}^{n} \left(v_j^- - v_{ij} \right)^2}$$

(4.10)

Step 6: At last, determine (the relative closeness of the overall preference or coefficient of closeness (CC)) the perfect solution for every alternative through utilizing Equation 4.10. Here, D_i^+ and D_i^- both are more than zero, then CC $\in$ (0, 1). We get from Equation 4.11,

$$CC_i = \frac{D_i^-}{D_i^+ + D_i^-} \quad \text{for } i = 1, 2, \ldots, m$$

(4.11)

Step 7: Finally, rank the options/alternatives in plunging order of preferences as indicated by the closeness coefficient (CC). The larger the CC, the betters the options/alternatives compared to others.

4.3 RESULTS AND DISCUSSION

4.3.1 Physical and Mechanical Characterizations

Figure 4.5 shows the influence of SiC-Gr on the voids content of different fillers (0, 2, 4, and 6 wt%) of Zinc aluminum (ZA27) alloy composite. It is observed that the density of fabricated composites shows a decreasing trend. The density follows the trend ZSG-4 > ZSG-2 > ZSG-0 > ZSG-6, while the voids content follows ZSG-4 < ZSG-0 < ZSG-6 < ZSG-2 across the formulation. It infers that the ZSG-4 alloy composite exhibits maximum density and moderate density value rise to minimum voids presence. This may be possible only when there is better synergy between complementary ceramic particulate and other ingredients of hybrid composites. It means that there are better interfacial adhesions or wettability between matrix and reinforcement that decrease the probability of voids formations, thereby making the matrix stiffer and denser. Therefore, the results of the voids content of prepared composite materials are evaluated from both the densities (such as theoretical and experimental). Voids of alloy composites are slowly improving by an increment in filler content. The consequences of voids depict the negative impact on the strength of materials and wear rate of composite.

The effect of SiC-Gr on the hardness of ZA27 alloy composites is shown in Figure 4.6. The hardness is enhanced with increment in reinforcement. The 6 wt% SiC-Gr of ZA27 alloy composites shows a higher hardness when contrasted to the base alloy composite. Therefore, the hardness of alloy composite growth is at the rate of 23%, 38%, and 60% higher base alloy. Therefore, the results of the hardness

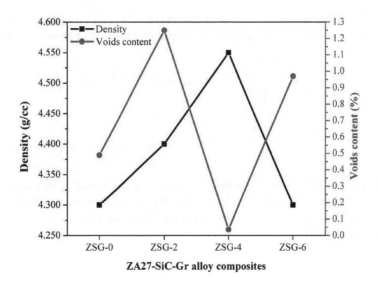

FIGURE 4.5   Density and voids content.

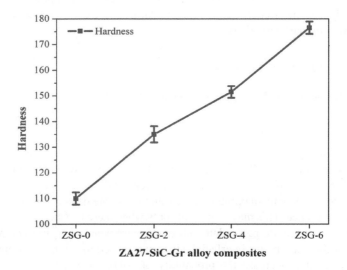

FIGURE 4.6   Hardness of the alloy composites.

of composite are increased because of the impendence of metallic hard particulate in base material matrix and load moved competency of the matrix toward the side of the filler was enhanced. The result outcomes have a great concurrence with the impact of mechanical properties on SiC–Gr filled ZA27 alloy composite [13,24].

The impact of tensile strength on SiC–Gr reinforced filled ZA27 alloy composite is depicted in Figure 4.7. It was estimated by utilizing the computerized UTM machine. It was concluded that the tensile strength results improve with increment in reinforcement. The strength of an alloy composite increases due to increased hardness and

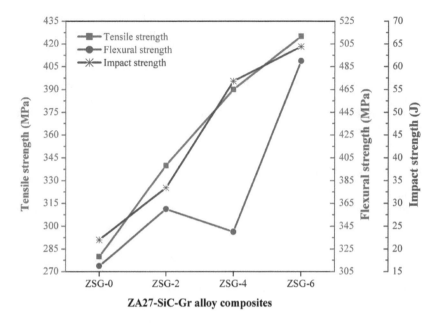

FIGURE 4.7 Tensile strength, flexural strength, and impact strength of the alloy composites.

dispersion of strength, and another reason is that the specimen of tensile strength increases with a diminished percentage of elongation [25]. It can see from the chart that it is 280 MPa at 0 wt% SiC/Gr particulate reinforcement. Further expansion of 2 wt% SiC/Gr content of strength of the composite is 340 MPa. Again, the addition of 4 wt% SiC/Gr and 6 wt% SiC/Gr content of composite strength are 390 and 425 MPa. A similar finding is also reported by Ramnath et al. [26]. This might be attributed to that most grounded holding among base material matrix and filler are playing a significant contribution in improving the specimen strength of the composite.

The variation of FS on SiC/Gr filled ZA27 alloy composites is employed in Figure 4.7. It is seen from the chart that the FS of the composite specimen enhanced to increment in filler content up to 2% by addition of weight percentage and then reduced with further reinforcing content. This may be an improper distribution particle and weak bonding generated because of improper casting. It concludes that the strength of the composite dependent on the overall impact of the size of particular reinforced particles, filler content/interlink bonding strength of the matrix, and impendence of voids content. The result outcomes stated by Baradeswaran et al. [8] studies show that the strength (flexural specimen) of composites improves with increment in particulate reinforcement in the alloy matrix.

The effect on the impact strength of a composite by the addition of SiC/Gr particles is employed in Figure 4.7. It is seen from the chart that the strength of composite specimen achieved by Izod experiment increases the wet.% of reinforcement. This strength is 22 J at 0 wt% SiC/Gr content. Again, the expansion of 2, 4, and 6 wt% of the reinforced content of impact strength of the composite is 33.4, 56.8, and 64.4 J. The maximum increase in impact strength of composite compared to base alloy was less than 19%. The same findings are stated by Kumar et al. [19] manifested that the impact strength improved because of ductile fracture in material and excellent holding reinforcement and matrix. This might be credited that the impact strength is superior at 6 wt% SiC/Gr because of the cracking in the material deposited at the energy of plastic deformation. The fabricated composites have extraordinary resistance ability because of increment in impact strength [27]. Kumar et al. [13] detailed a comparable trend to enhance reinforcement with increment in the impact strength of composite because of the temperature effect and the molecule size.

4.3.2 TAGUCHI ANALYSIS

The Taguchi technique is implemented in the present fabricated composite; the level of factors is selected on the steady-state test condition. Table 4.2 shows the scheme of the experimental simulation, and Table 4.3 shows the ranking order of input control factors. Figure 4.8 shows the main factor plot for S/N ratio, and Table 4.4 shows the result of ANOVA analysis and Mini tab 16 software are utilized for methodology-based computations.

In these investigations, L_{16} symmetrical collection DOE is implanted by having four factors and four levels. The entire mean for S/N proportions for the wear rate of the composite was acquired to be 77 dB. An entire sequence of the significance of control parameters in the diminishing rate of composite, i.e., sliding distance > normal load > reinforcement > sliding velocity [28–30]. ANOVA was utilized to observe the % share of control factors in calculating response, i.e., wear rate of composite.

TABLE 4.2
Taguchi L_{16} Orthogonal Array for Experimental Runs

Level	Normal Load	Reinforcement Content	Sliding Distance	Sliding Velocity	SWR	S/N Ratio (dB)
1	10	0	700	1	0.000745666	62.5491
2	10	2	1400	1.25	0.000189801	74.4340
3	10	4	2100	1.5	0.000293945	70.6347
4	10	6	2800	1.75	0.000111857	79.0268
5	20	0	1400	1.5	3.88368E-05	88.2151
6	20	2	700	1.75	0.000253068	71.9352
7	20	4	2800	1	0.00012204	78.2700
8	20	6	2100	1.25	9.05508E-05	80.8622
9	30	0	2100	1.75	3.45216E-05	89.2382
10	30	2	2800	1.5	0.000173985	75.1898
11	30	4	700	1.25	0.00015747	76.0560
12	30	6	1400	1	0.000127836	77.8669
13	40	0	2800	1.25	7.96154E-05	81.9801
14	40	2	2100	1	5.27226E-05	85.5601
15	40	4	1400	1.75	0.000181091	74.8421
16	40	6	700	1.5	0.000239693	72.4069

TABLE 4.3

Ranking Order of Significance of Control Factors

Levels	A (Load)	B (Reinforcement)	C (Sliding Distance)	D (Sliding Velocity)
1	71.66	80.50	70.74	76.06
2	79.82	76.78	78.84	78.33
3	79.59	74.95	81.57	76.61
4	78.70	77.54	78.62	78.76
Delta	8.16	5.54	10.84	2.70
Rank	2	3	1	4

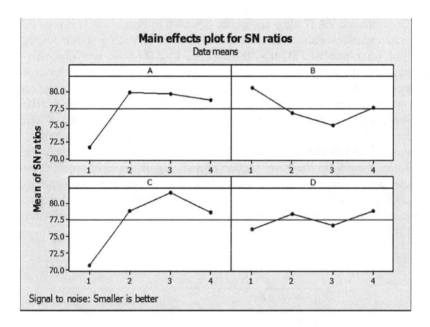

FIGURE 4.8 (A–D) Impact of control factors on the specific wear rate of alloy composites.

TABLE 4.4

ANOVA for Specific Wear Rate

Source	DF	Seq SS	Adj SS	Adj MS	F	P	P (%)
Normal load (L)	4	1400.73	1326.63	330.37	6.97	0.017	31.80
Filler content (F)	4	571.05	448.73	110.97	3.05	0.186	13.99
Sliding velocity (V)	4	551.11	595.85	147.62	3.68	0.111	11.54
Sliding distance (D)	4	1472.82	1372.89	342.58	7.18	0.015	33.52
Error	8	402.35	444.35	54.46			9.15
Total	24	4393.32					100

S = 8.35242; R-Sq = 87.87%; R-Sq(adj) = 67.64%.

In this case, the level of significance (represented by *p*) of 5% (i.e., the level of confidence of 95%) is considered for the analysis; thus the lower magnitude of *p* signifies a higher contribution of that factor in the output response or percentage contribution (*P%*) may be calculated to understand the same fact. Henceforth, the order of influence of input control factors and their contribution to the SWR are sliding distance (33.52%) > normal load (31.80%) > reinforcement (12.99%) > sliding velocity (12.54%). The sequence of significant control factors determines the entire mechanism of the wear system.

4.3.3   STEADY-STATE WEAR/FRICTION ANALYSIS (EFFECT OF SLIDING DISTANCE)

The influence of the wear rate of changing the sliding distance (700–1400 m) of SiC-Gr reinforced ZA27 alloy composite is shown in Figure 4.9. It is seen that the wear rate enhances with an increase in the distance of sliding of composites. The order of wear rate is ZSG-0 > ZSG-2 > ZSG-4 > ZSG-6. It was observed from the graph that the wear rate of composite depicts the lesser value of the lesser sliding distance (250 m), but beyond the increased 1400–2800 m sliding distance of specific wear was increased. Generally, the wear rate (unfilled alloy) is higher as related to developed alloy composites, but the higher content of alloy composite of wear rate was lower relative to another particulate alloy composite. It might be attributed that the wear rate of fabricated composites is improved with an enhancement in the sliding distance owing to the lower section of a sliding disk which is stable, then the particles were fractured and exhausted debris particles generated through the counter surface. It may be due to the fact that as the contact area at the sliding surface is enhanced with an increase in the interaction time and number of turns, simultaneously the specific wear rate is also increased [28].

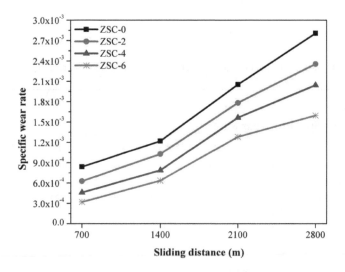

FIGURE 4.9   Effect of sliding distance on the specific wear rate of ZSG alloy composites.

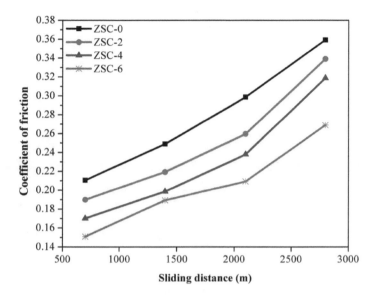

FIGURE 4.10   Effect of sliding distance on the coefficient of friction of ZSG alloy composites.

The impact of sliding distance (700–2800 m) on COF (μ) of SiC-Gr filled ZA 27 alloy composites is presented in Figure 4.10. The value of COF(μ) improves with increment in SiC-Gr (0–2 wt%) content of alloy composite. The COF value of composites shows the higher at 0 wt% filler content, and the lower value of COF shows at 2 wt% SiC/Gr powder reinforced ZA27 alloy composite. However, reduction in wear, damage of aluminum-lead alloy dependent on micrograph structure, specific sizes; distribution of grain shape; and lead are supportive to wear properties. It might be that the magnitude of COF was lower at a lower sliding distance for the ZSR alloy composite. The COF of the composite is slowly improved with an increase in sliding velocity for SiC-Gr filled ZA27 alloy composite. The COF(μ) is improved with the sliding distance due to the grain-growing mechanism and hardness increase in wear resistance [28,13].

4.3.4   Worn Surface Damage Analysis

The SEM micrograph of silicon carbide–graphite filled ZA27 alloy composites for sliding wear (steady-state condition; Taguchi DOE; L16 orthogonal array) at changing load (10–40 N) are presented in Figure 4.11a–d. It was evident that from graph Figure 4.11a, the wear rate (WR) of unfilled alloy composites was 0.000745666 mm³/Nm at level 1 (Table 4.2). This may be due to the delamination mechanism that may occur at lower applied normal load, i.e., 10 N and lower sliding velocity of 1 m/s due to small frictional heating between the interfaces of two materials. While delamination happens in lesser sliding speed and load that outcomes in the additional deficit of materials through delamination [29]. Figure 4.11b for 2% SiC-Gr filled ZA27 alloy composite demonstrates the specific wear rate, i.e., 0.000253068 mm³/Nm (exp. run 5, Table 4.2) with steady-state parameters (Sliding distance = 700 m;

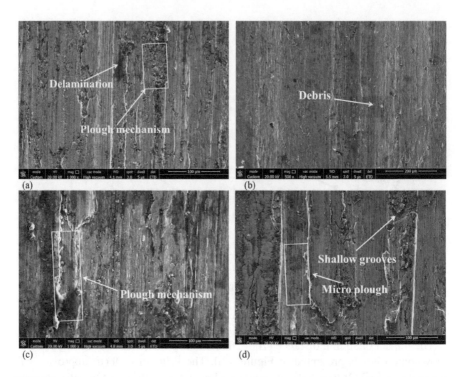

FIGURE 4.11 Worn surface micrographs of the alloy composites. (a) ZSG-0, (b) ZSG-2, (c) ZSG-4, and (d) ZSG-6.

Normal load= 20 N; Sliding speed = 1.75 m/s). The particle of debris (PD) is seen on the worn surface underneath the utilization of increment of load. (At the interface, heat is created due to the mellowing of the surface and impairment of wear particles [30].) ZA27 alloy with 4 wt% SiC-Gr filled composite material of the specific wear rates is 0.00015747 mm³/Nm (exp. run 11, Table 4.2) at steady-state parameters (Sliding distance = 700 m; Normal load = 30 N; Sliding speed = 1.25 m/s) are presented in Figure 4.11c. The ploughing mechanism generates because of the maximum content of plastic deformation [31]. The SEM micrograph of sliding wear in Figure 4.11d of 6 wt% of SiC-Gr filled ZA27 alloy composite indicates that the wear rate (SWR) is 0.000239693 mm³/Nm (exp. run 16; Table 4.2) at the steady state of sliding-wear parameters (high-sliding velocity = 1.5 m/s; normal load = 40 N; sliding distance = 700 m). In steady-state sliding tests, when the two flat pieces are rubbed together, they generate friction heats at the interface of the material surface because of the mellowing of surface, impairment of wear particles, and shallow grooves of the composite [32–34].

4.3.5 Hybrid AHP-TOPSIS Ranking Optimization

The ranking optimization of ZSG alloy composites is evaluated using hybrid AHP-TOPSIS methods as discussed in Section 2.6. The stepwise computations as per the AHP algorithm are as follows:

Step 1: The brainstorming sessions with subject experts and literature aid in gathering relevant information regarding investigated problem and may be arranged in the hierarchy structure as shown in Figure 4.12. Here, the decision goal of ranking material formulations that overall optimizes the performance remains at the top. The performance criteria as evaluated in the above sections remain at the center, and material alternatives like ZSG-0, ZSG-2, ZSG-4, ZSG-6 remain at the end. The implications of performance criteria on the decision goal are listed in Table 4.5.

Steps 2–3: In this step, a pairwise comparison matrix (shown in Table 4.6) is prepared in consultation with experts, thereafter relative weights are obtained. Thereafter, consistency verification of the obtained weights must be done to ensure robust ranking results. The maximum Eigenvalue $(\lambda_{max}) \sim 8.04$, Consistency Index (CI) = 0.0061, and Consistency Ratio (CR) = 0.00435, which is significantly $\ll 0.1$ (upper limit of

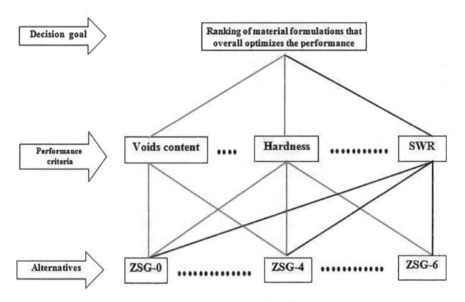

FIGURE 4.12 The hierarchy structure of the present problem.

TABLE 4.5
Performance Criteria and Their Implications

PC's No.	PC's	Implications of PC's
PC-1	Tensile strength (MPa); TS	Greater the better
PC-2	Flexural strength (MPa); FS	Lesser the better
PC-3	Impact strength (J); IS	Greater the better
PC-4	Hardness (HRB); HRB	Greater the better
PC-5	Specific wear rate (mm³/Nm); Wear	Lesser the better
PC-6	Density (g/cc); DEN	Lesser the better
PC-7	Voids content (%); VC	Lesser the better
PC-8	Coefficient of friction; COF	Lesser the better

TABLE 4.6

Pairwise Comparisons Matrix and Relative Weights

	PC-1	PC-2	PC-3	PC-4	PC-5	PC-6	PC-7	PC-8	Relative Weights
PC-1	1	1	1	1	2	2	3	2	0.147
PC-2	1	1	1	1	2	2	3	2	0.147
PC-3	1	1	1	1	2	2	3	2	0.147
PC-4	1	1	1	1	2	2	3	2	0.147
PC-5	1/2	1/2	1/2	1/2	1	1	2	1	0.053
PC-6	1/2	1/2	1/2	1/2	1	1	2	1	0.077
PC-7	1/3	1/3	1/3	1/3	1/2	1/2	1	1/2	0.141
PC-8	1/2	1/2	1/2	1/2	1	1	2	1	0.141
						Sum of relative weights:			1.00

acceptance of CR for consistency). From Table 4.6, it infers that order of priority of performance criteria's are Tensile strength ~ Flexural strength ~ Impact strength ~ Hardness (0.147) > Wear ~ Coefficient-of-friction (0.141) > Density (0.077) > Voids content (0.055).

Since the obtained relative weights are consistent, these may be used in the TOPSIS algorithm for ranking analysis of the alloy composites. The stepwise computations as per the TOPSIS algorithm is as follows:

Step 1: This step involves organizing the evaluated performance data in the form of a decision matrix as shown in Table 4.7.

Steps 2–5: The normalization of the decision matrix followed by the computation of closeness coefficients and ranking order is shown in Table 4.8.

It could be observed that the ranking of the alloy composites follows the order ZSG-6 > ZSG-4 > ZSG-2 > ZSG-0. The subjective analysis of the performance criteria discussed in Section 4.3.1–4.3.4 also reveals a similar ranking order. Thus, in this case, the ranking analysis is found to be in line with the subjective analysis. This may be due to a better and strong interface between ingredients with higher reinforcement of 6 wt% that keeps lowering as their proportion decreases [35–37].

TABLE 4.7

Evaluate Performance Data of Alloy Composites

	PC's							
	PC-1	PC-2	PC-3	PC-4	PC-5	PC-6	PC-7	PC-8
Materials Alternatives	TS	FS	IS	HRB	Wear	DEN	VC	COF
ZSG-0	280	310	22	110	2.8	4.3	0.49	0.19
ZSG-2	340	360	33.4	135	2.5	4.4	1.25	0.28
ZSG-4	390	340	56.8	151.5	1.8	4.55	0.036	0.35
ZSG-6	425	490	64.4	176.5	1.4	4.3	0.97	0.55

TABLE 4.8
Closeness Coefficient and Ranking of Alloy Composites

Materials Alternatives	D+	D−	CC	Ranking
ZSG-0	0.0495	0.1043	0.6783	4
ZSG-2	0.0332	0.1010	0.7528	3
ZSG-4	0.0316	0.0979	0.7556	2
ZSG-6	0.0002	0.0969	0.9980	1

4.4 CONCLUSION

The salient inferences made from the above investigation are as follows:

1. The mechanical characteristics of the alloy composites show improvement with reinforcement content, while density shows a reverse relation.
2. The specific wear rate and friction coefficient of the alloy composites observed an incremental trend with sliding distance irrespective of reinforcement content. The order of specific wear rate followed is ZSG-0 > ZSG-2 > ZSG-4 > ZSG-6 at any particular sliding distance.
3. The order of influence of control factors on specific wear is sliding distance > normal load > filler content > sliding velocity as obtained through the Taguchi method and validated with ANOVA analysis.
4. The study of wear data and the SEM micrograph of ZSG-6 alloy composites show superior wear performance compared to other composites; hence, it fits for bearing material application.
5. The maximum Eigen value $(\lambda_{max}) \sim 8.04$, Consistency Index (CI) = 0.0061, and Consistency Ratio (CR) = 0.00435, which is significantly $\ll 0.1$ (upper limit of acceptance of CR for consistency).
6. The order of priority of performance criteria is Tensile strength ~ Flexural strength ~ Impact strength ~ Hardness > Wear ~ COF > Density > Voids content.
7. The ranking of the alloy composites follows the order ZSG-6 > ZSG-4 > ZSG-2 > ZSG-0 and is found to be in line with the subjective analysis.

ACKNOWLEDGMENTS

The authors express their truthful gratitude to the Department of Mechanical Engineering of MNIT, Jaipur-302017, Rajasthan, India, for all kinds of financial and other miscellaneous infrastructural support. The authors also acknowledge the aid and facilities provided by the Advanced Research Lab for Tribology and Material Research Centre of the Institute for experimentation and characterization work.

REFERENCES

1. D.G. Wang, H.K. Peng, and J. Liu, Wear behavior and micro structural changes of SiCp-Al composite under lubricated sliding friction, *Wear* 184 (1995), pp. 187–192.

2. Z. Liu, G. Zu, and H. Luo, Influence of Mg addition on graphite particle distribution in the Al alloy matrix composites, *J Mater Sci Technol.* 26 (2010), pp. 244–250.

3. E. Karakulak, R. Yamanoglu, and U. Erten, Investigation of corrosion and mechanical properties of Al-Cu-SiC-xNi composite alloys, *Mater Des.* 59 (2014), pp. 33–37.

4. A. Kumar, A. Patnaik, and I.K Bhat, Tribology analysis of cobalt particulate filled Al7075 alloy for gear materials: A comparative study, *Silicon* 11 (2018), pp. 1295–1311.

5. E. Karthick, J. Mathai, and M. Joney, Processing, microstructure and mechanical properties of Al$_2$O$_3$ and SiC reinforced magnesium metal matrix hybrid composite, *Mater Today Proc.* 4 (2017), pp. 6750–6756.

6. K.K. Singh, S. Singha, and A.K. Srivastava, Comparison of wear and friction behavior of Aluminum metal matrix alloy (Al7075) and silicon carbide based Al metal matrix composite under dry condition at different sliding distance, *Mater Today Proc.* 4 (2017), pp. 8960–8970.

7. B.M. Girisha, K.R. Prakash, and B.M. Satisha, Need for optimization of graphite reinforcement in ZA-27 alloy composites for tribological application, *Mater Sci Eng A.* 530 (2011), pp. 382–388.

8. A. Baradiswaran, and A.E. Perumal, Influence of B4C on the tribological and mechanical properties of Al7075-B4C composites, *Int J Compos Part B* 54 (2013), pp. 146–152.

9. A. Kumar, and M. Kumar, Mechanical and dry sliding wear behavior of B4C and rice husk ash reinforced Al7075 alloy hybrid composite for armors applications by using Taguchi techniques, *Mater Today Proc.* 27 (2020), pp. 2617–2625.

10. M.M. Goudarzi, and P. Akhlaghi, Wear behavior of Al7252 alloy reinforced with micro metric and nano metric SiC particles, *Tribo Int.* 102 (2016), pp. 28–37.

11. F. Toptan, I. Kerti, and A.L. Rocha, Reciprocal dry sliding wear behavior of B$_4$Cp reinforced Al alloy matrix composites, *Wear* 290 (2012), pp. 74–78.

12. A. Kumar, and M. Kumar, Sliding wear performance of graphite reinforced AA6061 alloy for rotor drum/disk application, *Mater Today Proc.* 27 (2020), pp. 1972–1976.

13. A. Kumar, A. Patnaik, and I.K. Bhat, Investigation of nickel metal powder on tribological and mechanical propertiesofal 7075 alloy composites for gear material, *J Powder Metall.* 60 (2017), pp. 371–383.

14. S. Bhaskar, M. Kumar, and A. Patnaik, Application of hybrid AHP-TOPSIS technique in analyzing material performance of silicon carbide ceramic particulate reinforced AA2024 alloy composite, *Silicon.* 12 (2019), pp. 1075–1084.

15. G. Kranthi, and A. Satapathy, Evaluationandprediction of wear response of pine wood dust filled epoxy composites usin gneural computation. *Comput. Mater. Sci.* 49 (2010), pp. 609–614.

16. B.D. Agarwal, and L.J. Broutman, *Analysis and Performance of Fiber Composites*, 2nd edition, John Wiley & Sons, New York (1990).

17. A. Kumar, A. Patnaik, and I.K. Bhat, Sliding wear behavior of titanium metal powder filled aluminium alloy composites for gear application. *Int J Appl Mech Mater.* 877 (2018), pp. 118–136.

18. S.S. Kumar, M. Devaiah, and V. SeshuBai, SiCp/Al$_2$O$_3$ ceramic matrix composites prepared by directed oxidation of an aluminum alloy for wear resistance applications, *Ceram Int.* 38 (2012), pp. 1139–1147.

19. R. Kumar, K. Kiran, and V.S. Sreebalaji, Characterization of mechanical properties of aluminium/tungsten carbide composites, *Measurement* 102 (2017), pp. 142–149.

20. M. Kumar, Performance Assessment of Hybrid Composite Friction Materials : Effect of Ceramic, Organic and Inorganic Fibre Combinations, Ph.D Thesis, NIT Hamirpur, (2015).

21. M.C. Lin, C.C. Wang, M.S. Chen, and C.A. Chang, Using AHP and TOPSIS approaches in customer-driven product design process, *Comput. Ind.* 59 (2008), pp. 17–31.

22. R. Joshi, D.K. Banwet, and R. Shankar, A Delphi-AHP-TOPSIS based benchmarking framework for performance improvement of a cold chain, *Expert Syst Appl.* 38 (2011), pp. 10170–10182.

23. B.K. Satapathy, A. Majumdar, and B.S. Tomar, Optimal design of flyash filled composite friction materials using combined analytical hierarchy process and technique for order preference by similarity to ideal solutions approach, *Mater Des.* 31 (2010), pp. 1937–1944.

24. H. Abdizadeh, R, Ebrahimifard, and M.A. Baghchesara, Investigation of microstructure and mechanical properties of nanoMgO reinforced Al composites manufactured by stir casting and powder metallurgy methods: A comparative study, *Compos: Part B* 56(2014), pp. 217–221.

25. R. Kumar, K. Kiran, and V.S. Sreebalaji, Characterization of mechanical of aluminum/ tungsten carbide composite, *Measurement* 102(2017), pp. 142–149.

26. B. V. Ramnath, C. Elanchezhian, and M. Jaivignesh, Evaluation of mechanical properties of aluminum alloy–alumina–boron carbide metal matrix composites, *Mater Des.* 58(2014), pp. 332–338.

27. X. Li, C. Guo, and X. Liu, Impact behaviors of poly –latic acid based bio composite reinforced with unidirectional high strength magnesium alloy wires, *Prog Nat Sci.* 24 (2014), pp. 472–478.

28. A.K. Mondal, and S. Kumar, Dry sliding wear behavior of magnesium alloy based hybrid composites in the longitudinal direction, *Wear* 267 (2009), pp. 458–466.

29. H. Chi, L. Jiang, and G. Chen, Dry sliding friction and wear behavior of (TiB$_2$+h-BN)/2024Al composites, *Mater Des.* 87(2015), pp. 960–968.

30. A. Aherwar, A. Patnaik and C. Pruncu, Effect of B$_4$C and waste porcelain ceramic particulate reinforcements on mechanical and tribological characteristics of high strength AA7075 based hybrid composite, *J. Mater Res Tech.* 9(5) (2020), pp. 9882–9894.

31. C.S. Ramesh, R. Keshavamurthyand, and B.H. Channabasappa, Friction and wear behavior of Ni–P coated Si$_3$N$_4$ reinforced Al6061 composites, *Tribol Int.* 43 (2010), pp. 623–634.

32. J.C. Walker, W.M. Rain forth, and H. Jones, Lubricated sliding wear behavior of aluminum alloy composites, *Wear* 259 (2005), pp. 577–589.

33. H.R. Manohara, T.M. Chandrashekharaiah, K. Venkateswarlu, and S.A. Kori, Dry sliding wear response of A413 alloy: Influence of intermetallics and test parameters, *Tribol Int.* 51 (2012), pp. 54–60.

34. A Kumar, A Patnaik, and I.K. Bhat, Prametric analyasis of tribological for gear materials behaviour and mechanical study of cobalt metal powder filled Al-7075 alloy composites, *Mater Today: Proc.* 27 (2020), pp. 2787–2800.

35. S. Bhaskar, M. Kumar, and A. Patnaik, Silicon ceramic particulate reinforcement AA2024 alloy composite –Part I: Evaluation of mechanical and sliding tribology performance, *Silicon.* 12 (2019), pp. 843–865.

36. M. Kumar, Mechanical and sliding wear performance of AA356-Al2O3/SiC/Graphite alloy composite materials: Parametric and ranking optimization using Taguchi DoE and hybrid AHP-GRA method, *Silicon* (2020), doi: 10.1007/s12633-020-00544-9.

37. S. Dev, A. Aherwar, and A Patnaik, Material selection for automotive piston component using entropy –VIKOR method, *Silicon* 12 (2019), pp. 1557–1573.

5 Surface Texture Properties and Tribological Behavior of Additive Manufactured Parts

Binnur Sagbas
Yildiz Technical University

CONTENTS

DOI: 10.1201/9781003097082-5

5.1 INTRODUCTION

Tribology is a multidisciplinary science of interacting surfaces and includes several disciplines such as mechanical engineering, surface engineering (coatings, surface modification, surface analysis, and metrology), materials science, and chemical engineering. It is a comprehensive term that represents friction, wear, and lubrication subjects [1], which are very important phenomena in terms of service life of functional sliding parts, energy losses, environmental pollution, economy, and finally human life [2].

Friction is the resistance to the motion of sliding surfaces moving relative to each other. This resistance is highly influenced by surface topography, physics, and texture properties such as surface roughness, surface energy, wettability, etc. Therefore, it is highly affected by the manufacturing technique of the surface, post-processing, cleaning, coating, and surface modification. The friction force is expressed as $F = \mu W$, where W is the normal component of applied load for pressing the two sliding bodies and μ is the friction coefficient, which is highly dependent on texture properties of sliding surfaces. With the increase in surface roughness, bonding between the contact points of the surfaces occurs and the resistance of the surface increases, so a higher load is required for starting and maintaining the relative motion. Therefore, energy consumption and the cost of the sliding system also increase. During the motion, the input energy converts into frictional heating and dissipates into the system [3,4].

Surface properties of the sliding parts change by the frictional temperature rise. To decrease these effects, suitable lubricants in solid or liquid form are used in general. Lubrication is the most important way of reducing both friction and wear of the surfaces. Moreover, by the reduction of friction, energy consumption, and frictional heating also decreased. The effectiveness of lubrication is highly affected by the physical and chemical properties of the surface and lubricant. For fluid lubricants, lubrication film formation depends on the surface energy, wettability, roughness, and viscosity while powder size is important for solid lubricants [3]. Surface energy is the measure of impaired intermolecular bonds, which is generated during manufacturing, and it affects the wettability and lubrication of the surface. Surfaces with higher energy are very unstable and tend to lose their energy and become more stable. While a liquid drop is placed on a solid surface with high energy, the surface tends to interact with liquid and generate bonds. Therefore, the liquid drop spreads onto the surface. This is called wettability of the surface and is measured in terms of contact angle. Contact angle (θ) represents the angle between the liquid drop and the solid surface. While the contact angle is high, it refers that the wettability of the surface is low. Contact angle decreases with an increase in surface energy. The surface with 0° water contact angle is completely wettable. While the contact angle lower than 30° is defined as a super-hydrophilic surface, the surface with a contact angle between 30° and 90° is hydrophilic. Surface with a water contact angle between 90° and 120° is defined as hydrophobic and above 120° named as super-hydrophobic surfaces. A representative scheme of the surfaces is shown in Figure 5.1. Surface with high energy provides high wettability and better lubrication condition [5].

Three different lubrication regimes occur between sliding surfaces. These are boundary lubrication, mixed lubrication, and full film lubrication. In some references,

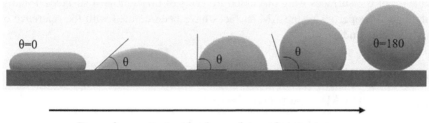

Increasing contact angle, decreasing surface energy

FIGURE 5.1 Representative drawing of different contact angles and surface energy relations.

elastohydrodynamic lubrication (EHL) regime is also counted as the fourth lubrication regime, which is considered as a subcategory of full film hydrodynamic lubrication in some other references. Surface texture properties, roughness values, and lubricant rheology have important effects on the lubrication regimes. While the arithmetic mean surface roughness value is higher than fluid film thickness, roughness peaks of the sliding surfaces will contact constantly, which is defined as boundary lubrication. In mixed lubrication, some of the roughness peaks contact each other, and the fluid film cannot provide full surface separation between the sliding surfaces. The full-film *lubrication regime* occurs when the fluid film completely separates the surfaces in relative motion. In that regime, the film thickness is much larger than the surface roughness. It is possible to control friction and wear by providing effective lubrication. Friction and wear decrease with achieving fluid film lubrication [1,4].

Wear is a complex phenomenon that causes gradual material loss from the sliding surfaces of functional solid bodies. It is affected by many factors such as surface properties, texture, hardness, material type, loading, lubrication, and temperature [6]. Different wear mechanisms such as adhesive, abrasive, erosive, corrosive, fretting, and fatigue wear aroused between sliding surfaces, and all of these wear mechanisms cause loss of functionality. It is stated in reference [7] that 23% of the total energy consumption of the word is caused by tribological contacts. Twenty percent of this consumption is caused by frictional losses where 3% of it is caused by remanufacturing and maintenance of worn parts. From the point of view of environmental concern, tribology also increases hazardous waste release by worn parts and used chemical lubricants. By taking advantage of technological development in materials science, lubrication technology, and the manufacturing industry for the generation of new functional surfaces, it is possible to decrease tribological losses and save about 40% of energy consumption in 15 years [7].

Additive manufacturing (AM) is a newly developing technology that provides an opportunity to generate complex-shaped functional parts with new surfaces. Because tribology has a great effect on the national and global economy, the characterization of newly generated surfaces and understanding their tribological characteristics are vitally important for developing long-lasting, energy-saving, and environmentally friendly functional system parts. This chapter aimed to define additive manufactured surfaces. After an overview of the AM technology, surface post-processing methods were defined. Surface texture properties and their importance on the tribological behavior of the surfaces were discussed in terms of different AM surfaces. Finally, tribological properties of the AM surfaces have been defined with the reference of experimental studies.

5.2   ADDITIVE MANUFACTURED SURFACES

5.2.1   ADDITIVE MANUFACTURING PROCESSES

AM is a newly developed and rapidly growing technology that provides an opportunity for generating three-dimensional (3D) functional parts by using 3D computer-aided design (CAD) data of the manufactured geometry. It was first developed in the 1980s and has been mostly used for rapid prototyping under the name of 3D printing. It is a layer-based manufacturing method that builds up 3D geometry without using any tools, dies, fixtures, or molds. Therefore, unlike subtractive and other conventional manufacturing methods, AM enables us to create complex geometries which provide design freedom without waste of materials [8].

AM processes start with the design of the geometry by any kind of CAD program and conversion of this data to the stereolithography file (STL). By using the related programs, the digital geometry gets sliced and layered according to the geometry, material, and desired specifications. Then the G codes are generated, and the processes start for building up 3D geometry [9]. The general steps for AM processes can be seen in Figure 5.2.

With the developing technology, AM techniques have become a promising manufacturing method for companies and it is started to be used for manufacturing near net-shaped functional parts for end users [10]. By providing an opportunity to generate freeform surfaces and lightweight bodies with different lattice structures, AM has been widely used in aerospace, defense, automotive, and medical industries [8]. A wide range of materials such as metals, polymers, ceramics, clays, and composites can be used in AM techniques [11]. Different AM methods have been developed for processing these materials, and the methods can be classified according to the type of material and form of the feedstock such as powder, bar, filament, resin, and slurry. Besides, it is possible to make a classification according to the heat source of the method such as laser, electron beam, UV, etc. The American Society for Testing and

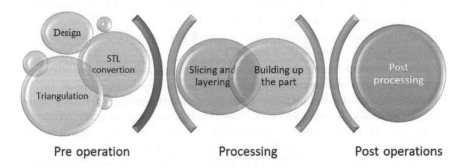

FIGURE 5.2   Steps of additive manufacturing processes.

Materials (ASTM) group, ASTMF42-Additive Manufacturing committee has classified AM methods in seven categories. These categories are vat photopolymerization, powder bed fusion (PBF), material extrusion, material jetting, binder jetting, directed energy deposition, and sheet lamination [9] listed in Table 5.1.

However, each of the AM methods has its own process parameters and different properties, and the factors that affect the surface quality of the AM parts are quite similar. Two basic ways are used for achieving the desired surface integrity of the AM parts. These ways are process parameter optimization and post-processing. Factors that affect the surface integrity of the AM parts are build orientation [13], layer thickness, support structures, laser power, road width, and speed [14–16]. In most of the AM methods, support structures are necessary for building up geometries with overhang. To removing these structures, cleaning is required. However, after cleaning, support structures leave marks and roughness which cause uneven surfaces with

TABLE 5.1
Classification of Additive Manufacturing Methods [8,12]

Classification	Method Name	Material	Description
VAT photopolymerization	Stereolithography Digital light processing	Photopolymers Ceramics	Uses UV light for selectively curing the photosensitive liquid polymer layer by layer in vat
Material extrusion	Fused deposition modeling	Thermoplastic polymers Sand	Builds up the 3D layered geometry by mechanically extruding the molten material
Powder bed fusion	Selective laser melting Selective laser sintering Direct metal laser sintering Electron beam melting Multijet fusion	Metals Polymers Ceramics	Uses energy beam to fuse the powders together in accordance with 3D designed geometry
Material jetting	Polyjet Inkjet Thermojet	Photopolymers Wax	Directly deposits wax or photopolymer droplets according to the desired geometry and uses UV light for curing the material.
Binder jetting	Inkjet	Metals Polymers Ceramic	Deposits binding agents onto the material powder, selectively according to the designed geometry
Sheet lamination	Ultrasonic consolidation Laminated object manufacturing	Metals Ceramics Hybrids	Generates 3D geometry by cutting sheet form of the material which is adhesively bonded in accordance with CAD model data
Direct energy deposition	Laser engineered net shape Direct metal deposition Direct light fabrication 3D laser cladding	Metals	Powder or wire form of the material is directly fed into the melt pool of the substrate

low quality. In Figure 5.3, the SLM manufactured Ti6Al4V surface can be seen after trimming support structures without any post-cleaning operations.

In Figure 5.4, the design of a part that contains inclined and overhang sections and FDM manufactured from the part that contains support structures can be seen.

Because of the layer-based nature of the AM methods, the surface quality of the manufactured part is highly affected by the thickness of each layer. To achieve better surface quality, the layer thickness must be small that causes longer manufacturing time and so higher cost. With the increase in layer thickness, surface roughness also increases [17]. This effect is prominent, especially on the inclined edges, circular geometries, and holes. It results from nonsmooth surfaces, which look like stair-stepping and is called "staircase effect" [14]. It emerges by slippage between layers of inclined surfaces and depends on the inclination angle of the external surface and layer thickness [18].

The effect of layer thickness on the staircase effect can be seen in Figure 5.5. In (a) part of the figure layer, thickness is lower so the edge surface of the inclined part is smoother. In (b), the layer thickness is higher than (a) and the surface is rough because of the stair-stepping effect. Therefore, to obtain better surface quality, it is important to choose proper manufacturing parameters in accordance with the designed geometry.

FIGURE 5.3   SLM manufactured Ti6Al4V surface after trimming support structures.

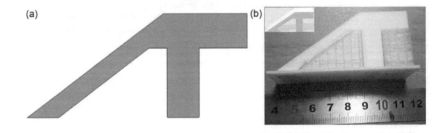

FIGURE 5.4   Design data of a workpiece with inclined and overhang sections (a), 3D geometry of the workpiece manufactured by FDM method, contains support structures just before the cleaning process (b).

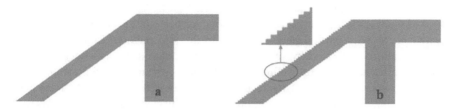

FIGURE 5.5 Effect of layer thickness on staircase effect with different layer thickness. Staircase effect with lower layer thickness (a), staircase effect with higher layer thickness (b).

FIGURE 5.6 Surface of SLM manufactured AlSi10Mg part

Another important factor that affects surface integrity of metal PBF AM parts is semi-molten powder grains that poorly stick to the outer surface of the part and remains spherical voids after being cleaned. These voids cause high surface roughness and high friction coefficient values. An example of surface texture and voids of SLM manufactured AlSi10Mg part surface can be seen in Figure 5.6. Post-processing techniques are applied to decrease these surface irregularities and roughness values.

5.2.2 Surface Post-Processing of AM Parts

Additive manufacturing provides a great opportunity for generating intricate geometry near net-shaped functional parts. However, surface integrity that affects the performance of these parts has not been at the desired value yet. The term surface integrity refers to the properties of surface and near-surface regions such as roughness, hardness, residual stress, oxidation, phase transformations, and pores. Low surface integrity represents low functional performance, especially from a tribological point of view [3]. Therefore, the improvement of surface quality is vitally important for enhancing tribological properties of the AM parts. The layer-based nature of the AM methods is a common reason for poor surface quality, although the surface quality of the built parts differs according to the AM method. Undesired surface texture features generated by semi-molten powder, stair-stepping effect, and support structures cannot be fully eliminated by process parameter optimization [8]. At that point, surface post-processing techniques are required. There are different post-processing methods applied for the improvement of surface quality of AM parts such as abrasive blasting, chemical etching [19], shot peening, polishing, machining, and laser-assisted finishing [20–23]. The list of the most commonly used post-processing methods can be seen in Table 5.2.

In this chapter, the most effective post-processing methods both for reducing surface integrity and enhancing tribological properties of the AM surfaces are discussed.

5.2.2.1 Abrasive Blasting

Abrasive blasting is a mechanical surface processing method that removes material and changes the texture of the surface by shooting the abrasives with high pressure [20]. Abrasive particles may scatter or rotate while they collide on the workpiece surface and generate an indentation or crater for ductile and brittle materials surfaces, respectively. The level of the abrasive impact velocity is a very important parameter for the process. Abrasive particles will only rotate, not scatter when the velocity is lower than the critical value and material removal by cutting cannot be achieved because of the inadequate energy, just plastic fatigue damage occurs. If the

TABLE 5.2

Post-Processing Methods for Additive Manufactured Parts

Material	Post-processing Method	Additive Manufacturing Method
Polymer	Abrasive blasting (sand, glass bead)	SLS, MJF [13], FDM [24]
	Chemical post-processing	FDM [25], SLS [19]
	Acetone bath	FDM [24]
	Ultrasonic-assisted Machining	FDM [26]
	Electroplating	FDM [27]
	Laser-assisted finishing (laser remelting, micromachining, polishing)	FDM [24,28]
	Magnetic field-assisted finishing (MFAF)	SLS [29]
	Precision grinding	SLS [29]
Metal	Abrasive blasting	SLM/DMLS [20]
	Shot peening	SLM/DMLS [5]
	Vibro finishing	SLM [30]
	Chemical post-processing	SLM [31]
	Laser-assisted finishing (laser remelting, micromachining, polishing)	SLM [32,33]
	Polishing (chemical, optical electropolishing)	SLM/DMLS [34]
Ceramic	Laser-assisted finishing	[21]
	Precision grinding	SLS [29]

impact velocity is high enough, the abrasives scatter, rotate, and create a crater onto the surface [35]. Besides impact velocity, abrasive type and geometry and feeding media of these abrasives to the surface are also important factors that affect the final surface texture and tribological behavior of the part. Different types of abrasives such as silica sand, aluminum pellets, copper shots, glass beads, steel grits, sodium bicarbonate, and even agricultural materials (olive pit) can be used as abrasive in the blasting process. The stream of these abrasive particles is fed to the surface in different media such as pressured air, fluid, or centrifugal force. The blasting process takes special names according to the abrasive type and feeding media. Such as wet blasting, bead blasting, wheel blasting, hydro-blasting, micro-abrasive blasting, dry-ice blasting, and vacuum blasting. Sandblasting [29], glass bead blasting [13], and steel grit blasting are generally used for polymer and metal additive manufactured surfaces [20]. The principle of abrasive blasting and images of SLM manufactured, abrasive-blasted AlSi10Mg surface can be seen in Figure 5.7.

5.2.2.2 Shot Peening

Shot peening is a mechanical surface post-processing method that causes cold plastic deformation of the surface bombarded with small spherical shots. Metal, ceramic, or glass spherical shots act like a small hammer and generate compressive stresses by imparting small dimples on the surface without any material removal. Near below the surface, compressed grains of the material tend to come back to their original position that causes high residual stress. Repeated shots generate a uniform surface layer of residual stress which increases mechanical properties, surface hardness, and

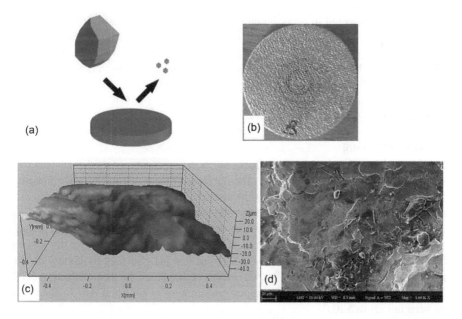

FIGURE 5.7 Principle of abrasive blasting (a); metal bead blasted, SLM manufactured AlSi10Mg surface (b); 3D texture view of the abrasive blasted AlSi10Mg surface (c); and SEM image of the AlSi10Mg surface (d) [20].

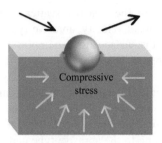

FIGURE 5.8   Schematic description of the shot-peening process

fatigue strength by work hardening. A schematic description of the shot peening process can be seen in Figure 5.8 [3,36,37].

During AM processes, especially in metal PBF processes, residual thermal stress and micro surface cracks, pores, or some other surface integrity problems arise because of the nonuniform temperature gradients and high cooling rates. These surface integrity problems decrease mechanical properties, load-carrying capacity, tribological properties, and service life of the functional parts. The shot peening process decreases surface pores, cracks, and crack propagation by severe plastic deformation of the surface. Peening exposure time, impact angle, shot size and type, shot velocity, and distance between nozzle and surface are the parameters that affect the performance of the shot peening process. Exposure time affects the intensity of the peening and determines the saturation of the surface. Shot material and its diameter have to be selected in accordance with the workpiece material. If the surface is harder than shot material, the shot will be deformed and the desired peening effect cannot be achieved. Moreover, fractured shots may cause material removal and damage the surface. It is preferred to apply the shots perpendicular to the surface. Angles smaller than 45° cannot provide sufficient deformation [38,39]. The principle of shot peening and images of SLM manufactured, shot-peened AlSi10Mg surface can be seen in Figure 5.9.

5.2.2.3   Polishing

Polishing is another type of mechanical surface processing method that works by the material removal principle from the surface. Different from abrasive blasting, polishing pads on which abrasive grains stacked are used for removing micron size chips from the surface [40]. Because the AM surfaces are rough, mechanical polishing has applied a sequence of stages that decrease the surface roughness gradually by using pads with different grit sizes. Asperities and higher peaks of the surface texture are removed at each repetitive polishing step [41]. Surface roughness of AM parts highly decreased [20] and fatigue resistance increased by the polishing process [42]. Schematic description of the process and images of polished AM surfaces can be seen in Figure 5.10.

A comparative microscopic image can be seen in Figure 5.11. It can be seen that surface pores on additive manufactured AlSi10Mg surface were decreased after the shot peening process while scratches were detected after polishing.

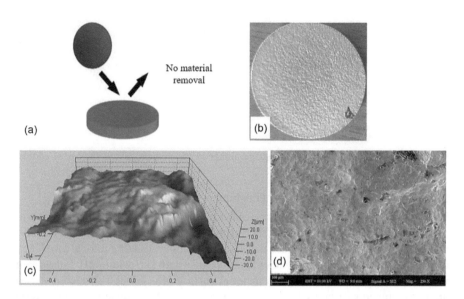

FIGURE 5.9 Principle of shot peening (a); metal shot-peened, SLM manufactured AlSi10Mg surface (b); 3D texture view of the shot-peened AlSi10Mg surface (c); and SEM image of the AlSi10Mg surface (d) [20].

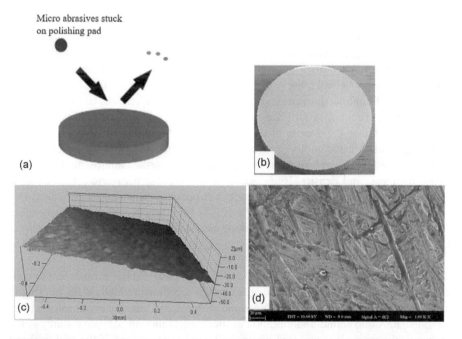

FIGURE 5.10 Principle of polishing (a); polished SLM manufactured AlSi10Mg surface (b); 3D texture view of the polished surface AlSi10Mg surface (c); and SEM image of the AlSi10Mg surface (d) [20].

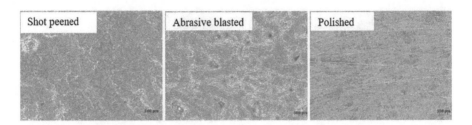

FIGURE 5.11   A comparative microscopic image of the post-processed surface of additive manufactured AlSi10Mg Parts

5.2.2.4   Laser-Assisted Surface Finishing

Laser-assisted surface finishing is one of the most widely used techniques by providing different laser-based processes such as micromachining and remelting. The laser micromachining process provides thermal energy by translating a long wavelength laser beam into a large spot diameter in accordance with kerf width [21]. Pulsed laser irradiation is used for melting the material surface [43]. Molten surface materials are redistributed by gravity and surface tension [30]. Micro-scaled molten material then solidifies and smoothens the part surface. The most important problem in laser micromachining is the heat-affected zone which causes residual stresses and metallurgical property changes. For decreasing these drawbacks, process parameters such as wavelength, spot size, intensity, and implementation of the laser have to be optimized [44].

Another effective laser-based surface quality improvement process for PBF technology is *laser remelting* which is applied to the top surface of the part during the selective laser melting process. It provides an opportunity to decrease surface roughness without any post-processing. By melting the top surface of the part, it fills the voids and valleys between the peaks [10] and smoothens the surface texture by the application of continuous wave laser radiation [43]. It is referred to as "surface shallow melting" mechanism [10]. Laser remelting improves surface quality by about 90% and increases the density of the SLM manufactured part by about 100% [45]. With rapid melting and cooling process, chemical transformations occur in the near-surface zone [46]. Therefore, laser remelting changes surface microstructure, chemistry, hardness, and density. By selecting proper process parameters such as laser power, scan speed, and laser wavelength, it is possible to increase mechanical and tribological properties of the AM surfaces [47–49]. Laser remelted, SLM manufactured Ti6Al4V surface and as-built surface can be seen in Figure 5.12.

5.2.2.5   Ultrasonic-Assisted Machining

Ultrasonic-assisted machining (UAM), which uses piezoelectric transducers to convert electrical energy into mechanical energy, is a promising method for improving the surface quality of AM parts. Ultrasonic piezoelectric transducers transform electrical signals into sound and vibration as mechanical motion. Abrasive slurry is fed between the workpiece and tool while the vibration frequency of the tool is above 20 KHz [50]. The process does not require electric conductivity, and it is nonthermal and does not cause chemical reactions and corrosion threats. By having these

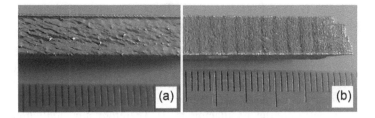

FIGURE 5.12   SLM manufactured Ti6Al4V surface; laser remelted (a), as build (b).

superiorities, besides conventional machining applications, UAM is used for improving FDM manufactured polymer parts surface quality [51].

5.2.2.6   Precision Grinding

Precision grinding is used for hard and brittle materials with relatively small surfaces to increase surface texture properties and form accuracy. According to the workpiece geometry, it can be applied in the form of horizontal, cylindrical, internal, and centerless grinding with the abrasive material of aluminum oxide, zirconia alumina, and silicon carbide. The method has been used for surface post-processing of SLM manufactured PA12 parts, and it is concluded that the precision grinding achieved a decrease in surface roughness from 15 to 2.85 μm Ra. Moreover, it decreased the surface hardness by releasing residual stresses and decreasing wear resistance of the surface [29].

5.2.2.7   Magnetic Field-Assisted Finishing

Magnetic field-assisted finishing (MFAF) uses a homogeneous particle mixture that contains abrasive and magnetic particles. The mixture is manipulated by the magnetic field to apply machining force to the workpiece [52]. Chips are removed from the workpiece surface by relative motion between particles and workpiece [29]. The method was used both for metal [53] and polymer additive manufactured surfaces. It is reported in the reference [29] that MFAF has decreased surface roughness and increased tribological performance of SLS manufactured PA12 part surfaces.

5.2.2.8   Chemical Post-Processing

Chemical post-processing is one of the most widely used finishing operations on both metal [31,54] and polymer [55]-based AM surfaces. The process can be applied under different media and special names such as chemical etching, acetone dipping, vapor smoothing, painting, and electroplating. Major superiorities of the chemical post-processes are that there is no geometrical restriction on the manufactured part for application of the post-processing. Moreover, there is no need to use any tool that contacts the workpiece surface, so it is possible to achieve better surface quality and dimensional accuracy [51]. The selection of proper chemicals for the workpiece material, concentration of the used solutions, portion of the chemicals, and duration of processing are important parameters that determine the effectiveness of the chemical post-processing.

5.2.3 Surface Texture Parameters and Their Effects on Tribological Behavior

Surface integrity and texture properties such as hardness, residual stress, roughness, and waviness of the functional machine parts have an important effect on defining their performance. Load-carrying capacity, tribological properties, corrosion resistance, and any other surface-related properties depend on the surface integrity and texture properties of the manufactured part. Therefore, it is vitally important to evaluate these properties and understand their effects on the tribological behavior of the functional parts. In this section, surface roughness parameters were defined and their effects on the tribological behavior of AM surfaces were discussed.

Manufacturing surfaces are built up from micrometer-scale deviations such as peaks, valleys, lays, and flaws. These deviations can be seen at high magnification even on the best-polished surfaces. While the two mating surfaces slide together, asperities tips contact first which cause a decrease in the real contact area and an increase in contact pressure. Plastic deformation and initial wear of the surface are seen on these asperity contacts [56]. The representative drawing of the first contact of the surfaces at macroscopic and microscopic levels can be seen in Figure 5.13.

Shape and distribution of the surface deviations have an important effect on friction, lubrication, and wear behavior of the machine parts. Each characteristic of the surface roughness is represented by a special parameter. These roughness characteristics can be measured by 2D (across a line) [57] or 3D (across an area) [58] techniques such as tactile mechanical profilometer and nontactile optical profilometer respectively. 2D surface roughness parameters have been used for decades for both quality control of the parts in the manufacturing industry and definition of surface texture properties in research studies. However, it is difficult to define surface texture properties of the functional parts by 2D line measurements because the surface of a workpiece is of a 3D (areal) nature. However, because of the complexity of 3D roughness, the characterization of 2D has been still used [59]. 3D areal surface parameters are derived by 2D line parameters. Therefore, in this section, the most important 2D surface roughness parameters (R) and their effects on tribological behavior of the surface were examined, which can be extended to 3D areal (S) parameters.

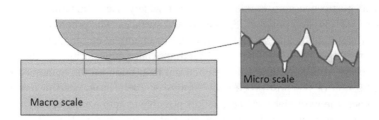

FIGURE 5.13 Representative drawing of the first contact of the real rough surfaces at macroscopic and microscopic level [1].

5.2.3.1 Arithmetic Mean Deviation

Arithmetic mean deviation (Ra) is one of the most widely used surface parameters and represents the arithmetric mean of the absolute deviations in $Z(x)$ ordinate, around the mean line, within the sampling length (Figure 5.14). The mean line is the least square mean line or generated by a standard filter and used for defining Ra by the following equation [60].

$$\mathrm{Ra} = \frac{1}{l}\int_0^l |Z(x)|\,dx \qquad (5.1)$$

Ra just defines the mean value of the surface roughness and does not differentiate between peaks and valleys. Therefore, it provides relatively weak information about the roughness character but it is an effective parameter for the determination of lubrication regime between the sliding surface [61]. Lubrication regime is defined by the relation between average surface roughness and film thickness. The ratio between minimum film thickness $h_{\min}$ and the composite roughness Ra, which is generally used for determining the distance between the frictional surface asperities, is expressed by λ parameter [62].

$$\lambda = \frac{h_{\min}}{\mathrm{Ra}} = \frac{h_{\min}}{\left[(\mathrm{Ra}_1)^2 + (\mathrm{Ra}_2)^2\right]^{1/2}} \qquad (5.2)$$

With the evaluation of λ correctly, the lubrication regime can be identified by the following ranges [61].

$0.1 < \lambda < 1$; boundary lubrication,
$1 \le \lambda \le 3$; mixed lubrication,
$\lambda > 3$; full film lubrication.

Schematic drawing of relation between the surface roughness and the film thickness can be seen in Figure 5.15.

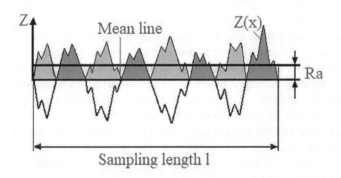

FIGURE 5.14 Representation of arithmetic mean deviation, Ra.

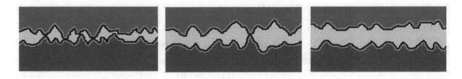

FIGURE 5.15 Schematic drawing of relation between surface roughness and film thickness [1]. (a) boundary lubrication, (b) mixed film lubrication, and (c) full film lubrication.

5.2.3.2 Root Mean Square Height

Root mean square height, Rq (RMS) is the standard deviation (σ) of the heights from the mean line, which represents the root mean square for $Z(x)$ within the sampling length (Figure 5.16). It is statistically meaningful and approximately 1.25 times higher than Ra if the heights are normally distributed [60,63].

$$\text{Rq} = \sqrt{\frac{1}{l}\int_0^l |Z^2(x)|\,dx} \tag{5.3}$$

Surface properties define the wear regimes by affecting the deformation mode of surface asperities, which is called plasticity index. It was first introduced by Greenwood and Williamson and can be calculated by the following formula:

$$\Psi = \frac{E^*}{H}\sqrt[2]{\frac{\sigma}{R}} \tag{5.4}$$

where Ψ is the plasticity index, E^* is Young's modulus, H is hardness of the softer material, R is radius of asperities, and σ is standard deviation of the asperity value which corresponds root mean square with other name Rq [64].

While $\Psi < 0.6$, asperities on the sliding surface would be deformed elastically. If the contact pressure is high enough, plastic deformation occurs. If $\Psi > 1.0$, asperities would be plastically deformed even if the contact pressure is low. While $0.6 < \Psi < 1.0$, deformation mode is in intermediate area and is not clear [65]. For understanding contact mechanics and determining tribological behavior of the functional surface, the mode of surface deformation is vitally important. Surface roughness or contact pressure has to be lowered to obtain elastic deformation mode. It is possible to design desired surface deformation mode by proper surface post-processing.

The model is generated for one processed, isotropic surfaces with a statistical assumption of Gaussian height distribution. However, after sliding motion of the surfaces, the initial symmetrical height distribution will turn into asymmetric height distribution. An example of this type of surface can be seen in Figure 5.17. The deformation at initial contact of the surfaces is plastic, and after the sliding period, it will be elastic and the texture properties of the surface will change. Therefore, Greenwood and Williamson model would not be enough for accurate calculations of the plasticity index. Different researchers proposed further models by likening the surface after a low wear process to two-process surfaces [64,65].

Dzierwa et al. compared the tribological properties of one-process (just processed by vapor blasting or lapping methods) and two-pen lapping methods) isotropic surface

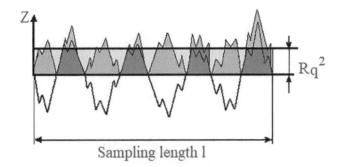

FIGURE 5.16 Representation of the root mean square height, Rq (RMS).

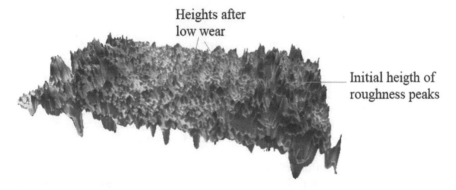

FIGURE 5.17 Asymmetric height distribution of a polymeric material surface after the low wear process.

with similar root-mean-square deviation values, and they reported that wear of two process samples was lower than one process sample's wear [66]. In another study, Pawlus et al. suggested a method for the determination of the texture properties (σ and R values) of a two-processed isotropic surface [64]. However, most of the engineering surfaces represent anisotropic surface characteristics. Therefore, the models proposed for isotropic surfaces are not suitable for determining the plasticity index of anisotropic surfaces. Newly developed models consider the degree of anisotropy γ and assume that all the roughness peaks shape is ellipsoidal.

$$\psi = \phi(\gamma)\frac{E'}{Re}\sqrt{\frac{\sigma}{Re}} \qquad (5.5)$$

Here, Re is an equivalent radius of roughness curvature, the geometric mean of the basic radii, $\varphi(\gamma)$ is a correction factor of near unity if $\gamma > 0.2$ [65]. However, these models developed for the determination of the cylinder linear surface characteristics, the last model may be the most suitable one for AM surfaces which are generally showing anisotropic properties because of the layer-based building principle. At that point, measurement of texture properties and defining their dimensions correctly gain

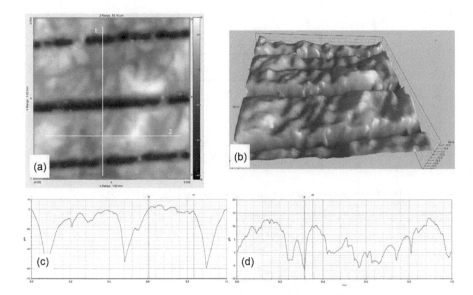

FIGURE 5.18 Effect of surface roughness measurement direction on FDM manufactured PLA part. Lines represented by the numbers 1 and 2 refer to measurement directions (a). 3D image of the surface (b). 2D surface profile obtained by the measurement direction 1 (c). 2D surface profile obtained by the measurement direction 2 (d).

importance for determining tribological properties of AM surfaces. It is important to take measurements that represent the whole surface homogeneously especially for anisotropic surfaces. For instance, a PLA surface manufactured by FDM can be seen in Figure 5.18. It is clear from the figure that measurement direction (represented by 1 and 2) highly affects the surface texture characterization results.

5.2.3.3 Maximum Height of Profile

Besides the mean values, it is important to characterize the maximum peak and valley heights that are affected by contamination, scratches, and measurement noise and have an important effect on the frictional behavior of the surface. Rv is the maximum depth of the valleys along the sampling length, while Rp is the maximum height value. Rz is the total height between top and deep points of the profile in sampling length and is determined by the absolute vertical distance between the Rv and Rp (Figure 5.19) [67].

$$Rz = Rv + Rp \qquad (5.6)$$

Height parameters just give information about the distribution of the deviations and their height values but they do not provide any data about height direction and horizontal features characteristics such as mean width of the profile elements, which highly affect contact pressure, lubrication, and friction properties of the surface. Therefore, for definition of the tribological behavior of the surface, further analysis and detailed characterization are needed.

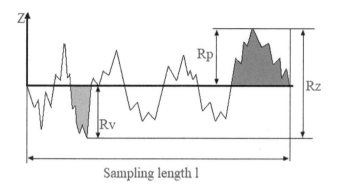

Sampling length l

FIGURE 5.19 Representation of the maximum height of profile.

5.2.3.4 Skewness

Skewness (Rsk) is a mean height direction parameter that gives information about the topography height distribution of the surface. It is the measure of profile symmetry about the mean line and is calculated by the ratio of the mean cube value of $Z(x)$ and the cube of the root mean square deviation within the sampling length. It is a unit-less parameter and represents different characteristic properties of the surface according to being negative, positive, and zero. Representation of the positive and negative skewness profile can be seen in Figure 5.20.

$$Rsk = \frac{1}{Rq^3}\left[\frac{1}{l}\int_0^l z^3(x)\,dx\right] \tag{5.7}$$

Rsk > 0: Deviation below the mean line
Rsk < 0: Deviation above the mean line
Rsk = 0: Symmetric against the mean line (normal distribution)

Skewness highly correlates with porosity, tribological properties, and load-carrying capacity of the surfaces. It defines the abrasion characteristic and sump of lubricant for bearing surfaces. Porous surfaces such as sintered and casted have a large value of negative skewness, which means a relatively higher number of deep valleys and rare peaks. Deep valleys provide better lubrication for sliding pairs by preventing lubricant and oil film. Positive skewed surfaces such as turned and grinded have a higher number of peaks which will contact counterbody first and simply worn out. Positive skewed surface cannot provide good lubricant retention by not having enough number of deep valleys. Therefore, for generating good bearing surfaces, negative skewness is a desired texture property [59,60].

5.2.3.5 Kurtosis

Kurtosis (Rku) defines the sharpness of the surface height distribution. It is the ratio of the fourth power of the deviations height values $Z(x)$ to the fourth power of the root mean square deviation within the sampling length.

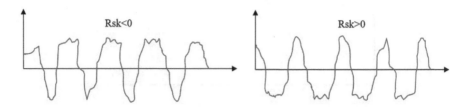

FIGURE 5.20 Representation of the skewness, Rsk.

$$Rku = \frac{1}{Rq^4}\left[\frac{1}{l}\int_0^1 z^4(x)\,dx\right] \qquad (5.8)$$

Kurtosis is a unit-less quantity like skewness and represents the spread of the height distribution.

Rku = 3: Normal (Gaussian) distribution
Rku > 3: The height distribution is sharp
Rku < 3: The height distribution is even

If a surface has a kurtosis value above 3, it means that the surface has sharp heights while low kurtosis denotes even heights. Figure 5.21 shows two profiles with low and high values of Rku.

Kurtosis just defines the tip geometry of surface deviations, and it does not recognize peaks and valleys. It is one of the important parameters for explaining the real contact between tribological pairs and their behavior in terms of wear and lubrication performance [59,60].

5.2.3.6 Mean Width of the Profile Elements

The mean width of the profile elements (RSm) indicates the mean value of the length Xs of profile elements that contain a profile peak and following valley along the sampling length (Figure 5.22). Height and spacing discrimination have to be specified for this parameter; otherwise, 10% of Rz for height discrimination and 1% of the sampling length for spacing discrimination defined as default [60].

$$RSm = \frac{1}{m}\sum_{i=1}^{m} Xsi \qquad (5.9)$$

RSm is an effective parameter for evaluating the horizontal dimensions of the parallel surface textures such as grooves and grains. The RSm parameter is important especially for the characterization of the layer-based AM surface texture and it is affected by layer thickness of the manufacturing process.

5.2.3.7 Material Ratio of the Profile

The material ratio of the profile is the ratio of bearing length to the evaluation length and is presented as percentage. Material ratio curve shows the amount of surface

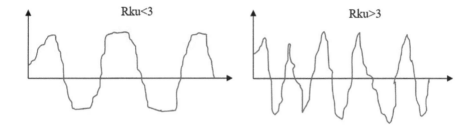

FIGURE 5.21 Representation of the kurtosis, Rku.

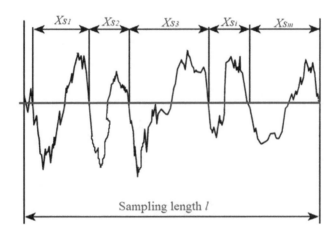

FIGURE 5.22 Representation of the mean width of the profile elements, RSm.

above a given height level for a profile, which is defined by drawing a line parallel to the mean line. If the slice level is at the highest peak, the ratio is 0% while it is 100% at the deepest valley. The material ratio curve is also known as the Abbott-Firestone curve (Figure 5.23) [60].

Rpk, Rk, Rvk, Mr1, and Mr2 parameters describe the wear characteristics in numerical terms by using a material ratio curve. These parameters were especially generated for the automotive industry for wear analysis of cylinder bores. They are frequently used to evaluate lubricity, friction, and abrasion behavior of the surfaces.

Rpk is reduced peak height that represents peaks of the surface. The high value of the Rpk refers to a surface that contains high peaks, which will be the first contact point of the surface with a small contact area and high contact stress. Therefore, this region of the surface will be the first worn part. Rpk also defines the load-bearing capacity of the surface. Rk is the Kernel roughness depth, which represents the roughness depth of the core region. This parameter is important especially for long-term sliding surfaces because it affects the performance and service life of the bearing surfaces. Rvk is the measure of valleys depth below the core region. It defines the lubricity of the surface. High Rvk values represent oil retention; moreover, these deep valleys serve as debris entrapment. Mr1 and Mr2 are the material ratios and refer to

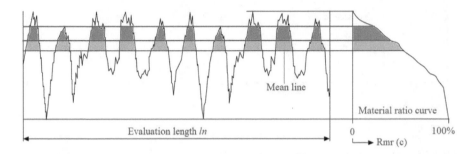

FIGURE 5.23   Material ratio curve of a profile.

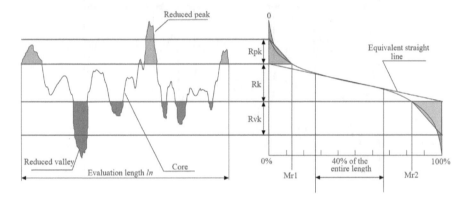

FIGURE 5.24   Bearing area curve.

the upper limit position and lower limit position of the roughness core respectively. All of these parameters can be seen in Figure 5.24 [67–69].

All of these parameters define tribological characteristics of an engineering surface such as from initial real contact pressure to achieved lubrication regimes. As mentioned above, the first contact of the sliding bodies occurs at the top surface of the peaks, which means that the real contact area is smaller than the nominal one. The representation of the first contact can be seen in Figure 5.25. In general engineering applications, the real contact area is about 1%–30% of the nominal contact area. The range depends on the surface texture of the sliding parts. Real contact pressure, which is calculated by using real area, is significantly higher than nominal pressure [56].

$$A_{reat} = \sum_{i=1}^{n} A_i \qquad (5.10)$$

$$P_{reat} = \frac{W}{A_{reat}} \qquad (5.11)$$

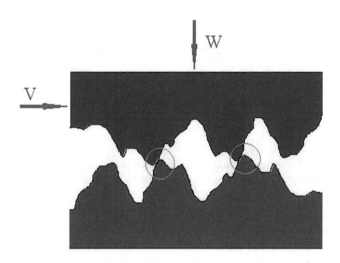

FIGURE 5.25 First contact points of the surfaces.

5.2.4 ADDITIVE MANUFACTURING SURFACES AND THEIR TRIBOLOGICAL PROPERTIES

Experimental investigations about 2D and 3D surface texture properties and tribological behavior of AM surfaces under different lubrication conditions are discussed here with reference to previous studies.

The tribological behavior of polymer-based FDM manufactured surfaces highly depends on process parameters and finishing operations. Adhesive, abrasive, and fatigue wear occurs by breaking of interfacial bonds which are affected by process parameters. There is a complex interaction between these parameters. As can be concluded from literature studies, the most effective parameter is layer thickness, which affects the bonding performance of the layers and so their wear resistance. While layer thickness decreases, heat transfer between the previous layer and the new layer increases, which causes local remelting and diffusion of the neighboring filaments. Therefore, fatigue, adhesion, and deformation resistance of the surface increase by strong bonding. Part orientation, infill density, shell number, raster angle, and raster gap also affect the bonding and so as to wear resistance [70–73]. Average roughness (Ra) of the FDM manufactured polymer surfaces was reported as from 4.57 to 25.3 μm while skewness values were negative without post-processing [16]. Therefore, it can be concluded that these surfaces provide better lubrication during sliding [63]. An example of a 2D roughness profile with high skewness and 3D optical profilometer image of the worn surface manufactured by FDM can be seen in Figure 5.26.

Krishna et al. [24] studied post-processing effects on the surface quality of the FDM manufactured parts. They applied acetone vapor smoothening, laser-assisted finishing, and shot blasting finishing methods to the ABS sample surfaces. They reported that acetone vapor smoothening provided the best surface quality, while laser-assisted finishing also achieved a significant decrease in surface roughness with

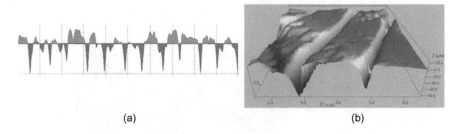

(a)   (b)

FIGURE 5.26 2D roughness profile of FDM manufactured PLA surface with high skewness (a) and 3D optical profilometer image of the worn surface manufactured by FDM (b).

implementing isotropic surface properties, and blasting time had a significant effect on final surface properties of FDM manufactured ABS parts surfaces. In another study [70], effect of FDM process parameters on wear behavior of ABS parts was investigated. It was reported that the wear rate of the parts increased by increasing raster angle and air gap while the wear rate decreased with a decrease in layer thickness and build orientation. The authors concluded that FDM process parameters highly affect the tribological properties of the manufactured part by leading to various microstructural changes. Sınmazçelik et al. [74] studied the erosive wear behavior of additive manufactured ABS surfaces. They built up the rectangular prismatic surface textures in different dimensions by the FDM method and reported that erosive wear resistance of the additive manufactured surface was highly affected by texture properties and erosion direction.

For PBF manufactured metal AM surfaces, Ra value can change in a wide range such as from 8 to 25 μm [20] depending on process parameters again. Texture properties, hardness, surface energy, and porosity of metal PBF parts are highly affected by the melting and cooling process which affects the microstructure of the part. Fine grains provide higher surface roughness and higher friction coefficient while they provide higher wear resistance by having a large number of grain boundaries and higher hardness values [75,76].

AlSi10Mg manufactured by PBF technology represented a higher friction coefficient and lower wear rate than casted AlSi10Mg. This was because of the smaller grain structure of the PBF manufactured part [77]. PBF manufactured Ti6Al4V represented higher surface hardness and better wear resistance than conventional Ti6Al4V. In the PBF process, α and α' hard phases were formed during rapid cooling which provides better wear resistance. Post-heat treatment of AM metal parts decreases wear resistance by the formation of coarse grains; however, post-heat treatment increases wear resistance of Ti6Al4V with the formation of protective oxide layer on the material surface [78]. Zhu et al. studied about tribological behavior of SLM manufactured 316L stainless steel. They compared AM surfaces performances with traditionally manufactured material surface, and they concluded that SLM manufactured samples had obviously lower friction coefficient and wear rate than traditionally manufactured samples [79].

Post-processing also has an important effect on the tribological behavior of metal AM parts. Shot peening, abrasive blasting, and polishing effects on tribological

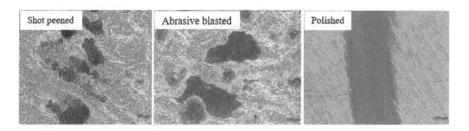

FIGURE 5.27 Microscopic images of the AM manufactured AlSi10Mg worn surfaces.

properties of DMLS manufactured AlSi10Mg parts were investigated in Ref. [20]. It was reported that minimum surface roughness and hardness were measured on the polished surface while maximum roughness and hardness values were recorded on the abrasive blasted surface. Negative skewness was recorded for shot-peened and polished surface, while it was positive and near to zero for abrasive blasted surface. As a result of generated surface properties, minimum friction coefficient and maximum wear factor were measured on polished surface while minimum wear factor was calculated for shot-peened surface. By having maximum roughness and hardness values, abrasive blasted surface wear factor was in medium value while the friction coefficient was highest. Roughness peaks on an abrasive blasted surface worn out first but it did not progress because of the high surface hardness.

Microscopic images of the worn surfaces of AM manufactured AlSi10Mg can be seen in Figure 5.27. The difference between the wear tracks reveals the effect of different post-processing.

5.3 CONCLUSION AND FUTURE REMARKS

Additive manufacturing is a rapidly growing technology, which has a wide range of application from architecture, aerospace to medical and defense industries. Desired properties of the AM products such as mechanical properties, dimensional accuracy, and surface quality alter according to the functional application area. Surface quality has an important effect on tribological properties and functionality of the manufactured products. Surface texture properties and roughness parameters can be improved by AM process parameters or post-processing applications. Although significant developments have been recorded in the AM technology, surface quality is still under consideration. Further research is needed to investigate proper surface post-processes for the products manufactured by different AM technologies with different materials. Furthermore, it is important for the development of measurement systems in parallel with manufacturing technology for the characterization of surface texture properties and their effects on functional and tribological properties of the AM parts.

REFERENCES

1. Sagbas, B., Biotribology of artificial hip joints, in *Advances in Tribology*, P.H. Darji, Editor. 2016: Croatia: InTech Open.
2. Bhushan, B., *Modern Tribology Handbook*. Vol. 2. 2001: Boca Raton, FL: CRC Press.

3. Kenneth, C.L. and O.A. Oyelayo, *Friction, Wear, Lubrication: A Textbook in Tribology,* K.C. Ludema and O.O. Ajayi, Editors. 2019: Boca Raton, FL: Taylor & Francis, CRC Press.

4. Bruce, R.W., *Handbook of Lubrication and Tribology, Volume II: Theory and Design.* Vol. 2. 2012: Boca Raton, FL: CRC press.

5. Cuie, W., *Surface Coating and Modification of Metallic Biomaterials.* 2015: Sawston: Woodhead Publishing.

6. Zhu, Y., J. Zou, and H.-Y. Yang, Wear performance of metal parts fabricated by selective laser melting: A literature review. *Journal of Zhejiang University-Science A,* 2018. **19**(2): pp. 95–110.

7. Holmberg, K. and A. Erdemir, Influence of tribology on global energy consumption, costs and emissions. *Friction,* 2017. **5**(3): pp. 263–284.

8. Gibson, Y., D. Rosen, and B. Stucker, *Additive manufacturing Technologies: 3D Printing, Rapid Prototyping, and Direct Digital Manufacturing.* Second Edtion. 2015: New York: Springer.

9. Amit, B. and B. Susmita, *Additive Manufacturing.* 2016: Boca Raton, FL: Taylor & Francis Group CRC Press.

10. Srivatsan, T. and T. Sudarshan, *Additive Manufacturing: Innovations, Advances, and Applications.* 2016: Boca Raton, FL: CRC Press.

11. Badiru, A.B., V.V. Valencia, and D. Liu, *Additive Manufacturing Handbook: Product Development for the Defense Industry.* 2017: Boca Raton, FL: CRC Press.

12. Subramanian, S.M. and M.S. Monica, *Handbook of of Sustainability in Additive Manufacturing.* Vol. 1. 2016: Singapore: Springer.

13. Sagbas, B., Effect of orientation angle on surface quality and dimensional accuracy of functional parts manufactured by multi jet fusion technology. *Journal of European Mechanical Science,* 2020. **4**(2): pp. 47–52.

14. Brandt, M., *Laser Additive Manufacturing: Materials, Design, Technologies, and Applications.* 2016: Sawston: Woodhead Publishing.

15. Pérez, M., et al., Surface quality enhancement of fused deposition modeling (FDM) printed samples based on the selection of critical printing parameters. *Materials (Basel, Switzerland),* 2018. **11**(8): pp. 1382.

16. Sagbas, B., Surface texture characterization and parameter optimization of fused deposition modelling process. *Düzce Üniversitesi Bilim ve Teknoloji Dergisi,* 2018. **6**(4): pp. 1028–1037.

17. Hafsa, M., et al., Study on surface roughness quality of FDM and MJM additive manufacturing model for implementation as investment casting sacrificial pattern. *Journal of Mechanical Engineering,* 2018. **5**(6): pp. 25–34.

18. Di Angelo, L., P. Di Stefano, and A. Marzola, Surface quality prediction in FDM additive manufacturing. *The International Journal of Advanced Manufacturing Technology,* 2017. **93**(9–12): pp. 3655–3662.

19. Crane, N.B., et al., Impact of chemical finishing on laser-sintered nylon 12 materials. *Additive Manufacturing,* 2017. **13**: pp. 149–155.

20. Sagbas, B., Post-processing effects on surface properties of direct metal laser sintered AlSi10Mg parts. *Metals and Materials International,* 2020. **26**(1): pp. 143–153.

21. Kumbhar, N. and A. Mulay, Post processing methods used to improve surface finish of products which are manufactured by additive manufacturing technologies: A review. *Journal of The Institution of Engineers: Series C,* 2018. **99**(4): pp. 481–487.

22. Barari, A., et al., On the surface quality of additive manufactured parts. *The International Journal of Advanced Manufacturing Technology,* 2017. **89**(5–8): pp. 1969–1974.

23. Mohammadi, M. and H. Asgari, Achieving low surface roughness AlSi10Mg_200C parts using direct metal laser sintering. *Additive Manufacturing,* 2018. **20**: pp. 23–32.

24. Krishna, A.V., et al., Influence of different post-processing methods on surface topography of fused deposition modelling samples. *Surface Topography: Metrology Properties*, 2020. **8**(1): p. 014001.

25. Valerga, A.P., et al., Impact of chemical post-processing in fused deposition modelling (FDM) on polylactic acid (PLA) surface quality and structure. *Polymers*, 2019. **11**(3): p. 566.

26. Maidin, S., M. Muhamad, and E. Pei. Experimental setup for ultrasonic-assisted desktop fused deposition modeling system. Applied Mechanics and Materials, 2015. 761: pp. 324–328.

27. Sugavaneswaran, M., P.M. Thomas, and A. Azad, Effect of electroplating on surface roughness and dimension of FDM parts at various build orientations. *FME Transactions*, 2019. **47**(4): pp. 880–886.

28. Kumbhar, N. and A. Mulay. Finishing of fused deposition modelling (FDM) printed parts by CO_2 laser. in *Proceedings of 6th International & 27th All India Manufacturing Technology, Design and Research Conference,* Pune, Maharashtra, India. 2016. pp. 16–18.

29. Guo, J., et al., Surface quality improvement of selective laser sintered polyamide 12 by precision grinding and magnetic field-assisted finishing. *Materials & Design (Basel, Switzerland)*, 2018. **138**: pp. 39–45.

30. Atzeni, E., et al., Performance assessment of a vibro-finishing technology for additively manufactured components. *Procedia CIRP*, 2020. **88**: pp. 427–432.

31. Łyczkowska, E., et al., Chemical polishing of scaffolds made of Ti–6Al–7Nb alloy by additive manufacturing. *Archives of Civil and Mechanical Engineering*, 2014. **14**(4): pp. 586–594.

32. Gora, W.S., et al., Enhancing surface finish of additively manufactured titanium and cobalt chrome elements using laser based finishing. *Physics Procedia*, 2016. **83**: pp. 258–263.

33. Worts, N., J. Jones, and J. Squier, Surface structure modification of additively manufactured titanium components via femtosecond laser micromachining. *Optics Communications*, 2019. **430**: pp. 352–357.

34. Tyagi, P., et al., Reducing the roughness of internal surface of an additive manufacturing produced 316 steel component by chempolishing and electropolishing. *Additive Manufacturing*, 2019. **25**: pp. 32–38.

35. Lin, C.-H. and C. Zhu, Chapter 3- Relevance of particle transport in surface deposition and cleaning, in *Developments in Surface Contamination and Cleaning.* Second Edition, R. Kohli and K.L. Mittal, Editors. 2008: William Andrew Publishing: Oxford. pp. 91–118.

36. Wu, Q., et al., Effect of shot peening on surface residual stress distribution of SiCp/2024Al. *Composites Part B: Engineering*, 2018. **154**: pp. 382–387.

37. Fedoryszyn, A., and P. Zyzak, Materials, Characteristics of the outer surface layer in casts subjected to shot blasting treatment. *Archives of Metallurgy*, 2010. **55**(3): pp. 813–818.

38. Avcu, E., Surface properties of AA7075 aluminium alloy shot peened under different peening parameters. *Acta Materialia Turcica*, 2017. **1**(1): pp. 3–10.

39. Wagner, L., *Shot Peening.* Vol. 8. 2003: Chichester: John Wiley & Sons.

40. Zhou, C., et al., Influence of colloidal abrasive size on material removal rate and surface finish in SiO_2 chemical mechanical polishing. *Tribology Transactions*, 2002. **45**(2): pp. 232–238.

41. Jin, S., et al., A gaussian process model-guided surface polishing process in additive manufacturing. *Journal of Manufacturing Science*, 2020. **142**(1).

42. Chan, K.S., et al., Fatigue life of titanium alloys fabricated by additive layer manufacturing techniques for dental implants. *Metallurgical Materials Transactions A*, 2013. **44**(2): pp. 1010–1022.

43. Juliana, d.S.S., J.S. Hans, and P. Wilhelm, Laser surface modification and polishing of additive manufactured metallic parts. *Procedia CIRP*, 2018. **74**: pp. 280–284.
44. Bhattacharyya, B. and B. Doloi, Chapter Seven - Micromachining processes, in *Modern Machining Technology*, B. Bhattacharyya and B. Doloi, Editors. 2020: London: Academic Press. pp. 593–673.
45. Yasa, E., J. Deckers, and J.P. Kruth, The investigation of the influence of laser re-melting on density, surface quality and microstructure of selective laser melting parts. *Rapid Prototyping Journal*, 2011. **17**(5): pp. 312–327.
46. Vaithilingam, J., et al., The effect of laser remelting on the surface chemistry of Ti6al4V components fabricated by selective laser melting. *Journal of Materials Processing Technology*, 2016. **232**: pp. 1–8.
47. Liu, B., B.-Q. Li, and Z. Li, Selective laser remelting of an additive layer manufacturing process on AlSi10Mg. *Results in Physics*, 2019. **12**: pp. 982–988.
48. Yu, Y., et al., The effects of laser remelting on the microstructure and performance of bainitic steel. *Metals*, 2019. **9**(8): p. 912.
49. Major, B., Chapter 7: Laser processing for surface modification by remelting and alloying of metallic systems, in *Materials Surface Processing by Directed Energy Techniques*, Y. Pauleau, Editor. 2006: London: Elsevier.
50. Wang, J., et al., Material removal during ultrasonic machining using smoothed particle hydrodynamics. *International Journal of Automation Technology*, 2013. **7**(6): pp. 614–620.
51. Singh, R. and J.P. Davim, *Additive Manufacturing: Applications and Innovations*. Manufacturing Design and Technology Series, J.P. Davim, Editor. 2018: Boca Raton, FL: CRC Press.
52. Zou, Y.H. and T. Shinmura, Mechanism of a magnetic field assisted finishing process using a magnet tool and magnetic particles. *Key Engineering Materials*, 2007. **339**: pp. 106–113.
53. Yamaguchi, H., O. Fergani, and P.-Y. Wu, Modification using magnetic field-assisted finishing of the surface roughness and residual stress of additively manufactured components. *CIRP Annals*, 2017. **66**(1): pp. 305–308.
54. Scherillo, F., Chemical surface finishing of AlSi10Mg components made by additive manufacturing. *Manufacturing Letters*, 2019. **19**: pp. 5–9.
55. Garg, A., A. Bhattacharya, and A. Batish, Chemical vapor treatment of ABS parts built by FDM: Analysis of surface finish and mechanical strength. *The International Journal of Advanced Manufacturing Technology*, 2017. **89**(5–8): pp. 2175–2191.
56. Bhushan, B., *Modern Tribology Handbook*, two volume set. 2000: Boca Raton, FL: CRC press.
57. ISO, ISO 4287:2010-07 Geometrical Product Specifications (GPS) - Surface texture: Profile method - Terms, definitions and surface texture parameters. 2010.
58. ISO, ISO 25178-2:2012 Geometrical product specifications (GPS) — Surface texture: Areal- Part 2: Terms, definitions and surface texture parameters. 2012.
59. Leach, R., *Characterisation of Areal Surface Texture*. 2013: Berlin Heidelberg: Springer.
60. Leach, R., *Fundamental Principles of Engineering Nanometrology*. 2014: Alpharetta, GA: Elsevier.
61. Mattei, L., et al., Lubrication and wear modelling of artificial hip joints: A review. *Tribology International*, 2011. **44**(5): pp. 532–549.
62. Di Puccio, F. and L. Mattei, Biotribology of artificial hip joints. *World Journal of Orthopedics*, 2015. **6**(1): p. 77.
63. Alsoufi, M.S. and A.E. Elsayed, Surface roughness quality and dimensional accuracy—a comprehensive analysis of 100% infill printed parts fabricated by a personal/desktop cost-effective FDM 3D printer. *Materials Sciences Applications*, 2018. **9**(1): pp. 11–40.

64. Pawlus, P., et al., Calculation of plasticity index of two-process surfaces. *Proceedings of the Institution of Mechanical Engineers, Part J: Journal of Engineering Tribology*, 2017. **231**(5): pp. 572–582.
65. Pawlus, P., W. Grabon, and D. Czach. Calculation of plasticity index of honed cylinder liner textures. in *Journal of Physics: Conference Series*. 2019. IOP Publishing.
66. Dzierwa, A., et al. The study of the tribological properties of one-process and two-process textures after vapour blasting and lapping using pin-on-disc tests. Key Engineering Materials, 2013. **527**: pp. 217–222.
67. Olympus. https://www.olympus-ims.com/en/metrology/surface-roughness-measurement-portal/parameters/ Access Date: 18.07.2020.
68. TaylorHobson. https://www.taylor-hobson.com/resource-center/faq/what-are-hybrid-parameters Access Date: 18.07.2020.
69. ISO, ISO 13565-2:1996 Geometrical Product Specifications (GPS) — Surface texture: Profile method; Surfaces having stratified functional properties — in Part 2: Height characterization using the linear material ratio curve 1996.
70. Mohamed, O.A., et al., Investigation on the tribological behavior and wear mechanism of parts processed by fused deposition additive manufacturing process. *Journal of Manufacturing Processes*, 2017. **29**: pp. 149–159.
71. Sood, A.K., et al., An investigation on sliding wear of FDM built parts. CIRP Journal of Manufacturing Science Technology, 2012. **5**(1): pp. 48–54.
72. Gurrala, P.K. and S.P. Regalla, Friction and wear rate characteristics of parts manufactured by fused deposition modelling process. *International Journal of Rapid Manufacturing*, 2017. **6**(4): pp. 245–261.
73. Srinivasan, R., et al., Influential analysis of fused deposition modeling process parameters on the wear behaviour of ABS parts. *Materials Today: Proceedings*, 2020.
74. Sınmazçelik, T., S. Fidan, and S. Ürgün, Effects of 3D printed surface texture on erosive wear. *Tribology International*, 2020. **144**: p. 106110.
75. Li, H., et al., Effect of process parameters on tribological performance of 316L stainless steel parts fabricated by selective laser melting. *Manufacturing Letters*, 2018. **16**: pp. 36–39.
76. Lorusso, M., Tribological and wear behavior of metal alloys produced by laser powder bed fusion (LPBF), in *Friction, Lubrication and Wear*, M.A. Chowdhury, Editor 2019: Crotaria: IntechOpen.
77. Lorusso, M., et al., Tribological Behavior of Aluminum Alloy AlSi10Mg-TiB2 Composites Produced by Direct Metal Laser Sintering (DMLS). *Journal of Materials Engineering and Performance*, 2016. **25**(8): pp. 3152–3160.
78. Zhu, Y., et al., Sliding wear of selective laser melting processed Ti6Al4V under boundary lubrication conditions. *Wear*, 2016. **368**: pp. 485–495.
79. Zhu, Y., et al., Tribology of selective laser melting processed parts: Stainless steel 316L under lubricated conditions. *Wear*, 2016. **350–351**: pp. 46–55.

6 Wear and Corrosion of Wind Turbines

Jyoti Menghani and Prajal Nandedwalia
SVNIT

CONTENTS

6.1 INTRODUCTION

The aim of clean energy future which avoids environmental pollution and decreases global warming has compelled researchers to divert toward wind power as a reliable and emission-free energy source available in nature; it has become the world's fastest-growing renewable energy source. Wind energy is one of the most important sustainable and renewable energy sources which can be converted into electrical energy for both industrial power supplies and household electricity consumption [1–4].

The total energy production share of wind energy is continuously increasing on a very rapid scale. Presently, the installation of wind turbines is at its peak and the focus is shifting from onshore (turbines that are located on land) to offshore (turbines

DOI: 10.1201/9781003097082-6

115

located in the ocean) locations. Figures 6.1 and 6.2 respectively represent the projection of wind energy growth worldwide with more development of offshore compared to onshore wind turbines [5,6].

In addition, installation and the fast-growing wind industry in more than 100 countries have given employment to more than one million people worldwide. The development of the cumulative wind installations throughout 35 years (1981–2016) has been plotted in Figure 6.3 [7].

Offshore wind turbines have various benefits which include advantages of location and environmental constraints inland or public concerns, higher wind speeds

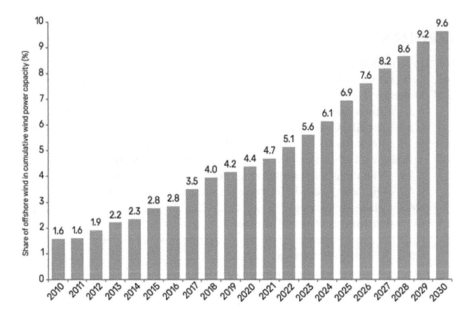

FIGURE 6.1　Global wind energy growth scenario.

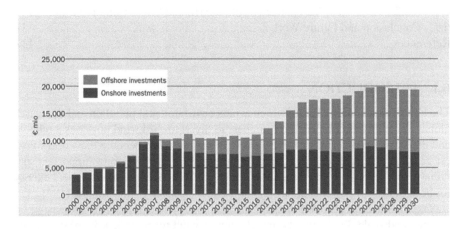

FIGURE 6.2　Development of onshore and offshore wind energy.

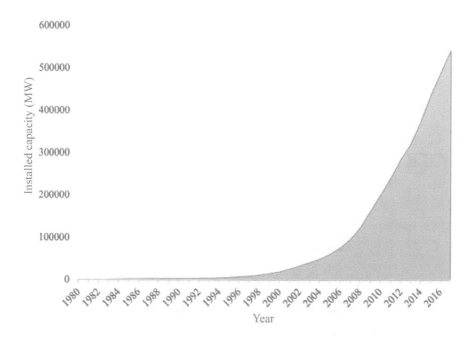

FIGURE 6.3   The global installed wind capacity from 1982 to 2017 [7].

and lower turbulence, and less visual and noise impacts; hence, most wind power installations have shifted to offshore sites. It is believed that around 2030 nearly 7.7% of overall electricity consumption in Europe will be covered by power production from offshore wind turbines by an installed power capacity of 66 GW [8].

6.2 COMPONENTS OF WIND TURBINES

The principle used in wind turbines is the conversion of kinetic wind energy (mechanical) to electrical components and is opposite to fan wherein electrical energy is converted to mechanical energy. Figure 6.4 shows the important components of wind turbines. Wind turbine rotor components consist of a set of blades attached to the rotor hub. The rotor is deflected by airflow creating force on blades which produce torque on the shaft which is connected to a gearbox and generator. The gearbox and generator are inside the nacelle and are located at the top end of the tower. The gearbox can increase rotation speeds. The common ratio is 90:1, input is 16.7 rpm and converted to @1500 rpm), which allows the generator to connect to the electric grid [2,10].

Modern wind turbines can be broadly classified into two distinct types, based on the orientation of the rotating axis: (1) horizontal axis wind turbines (HAWTs) (here the rotating axis of the wind turbine is horizontal, or parallel with the ground and is in the direction of the wind) and (2) vertical axis wind turbines (VAWTs) (rotational axis of the turbine stands vertical or perpendicular to the ground) [4,11,12]

Typically, the VAWT can be efficiently used in a residential area as it can generate more power at a relatively low speed. However, under the same operating conditions,

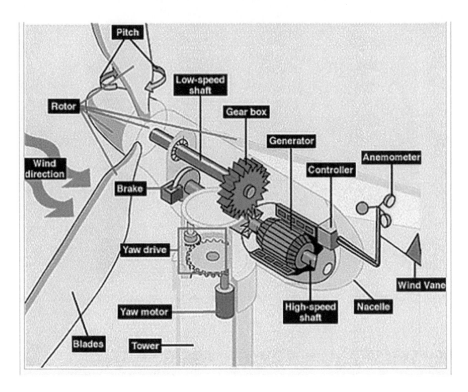

FIGURE 6.4   Components of wind turbine [9].

HAWT will generate more power since aerodynamic drag is less and higher power is generated due to the entire rotation of all blades. Table 6.1 gives the comparison of HAWT and VAWT.

6.3   MATERIALS USED FOR WIND TURBINE COMPONENTS

As shown in Figure 6.5, there is a combination of various types of materials in the construction of offshore wind turbines, including polymers, composites, concrete, and metallic alloys. The criteria for the selection of typical construction material depends on the size and part of the wind turbine and its location. Figure 6.5 also indicates the future trend of various materials used in wind turbines.

6.3.1   TOWER AND FOUNDATION

It provides structural support, and on it, nacelle and rotor blades stand. Typically, for a 5 MW turbine, the rotor has 120 t weight and the nacelle weight is 300 t. It constituent a major portion of the overall cost of an installation. The tower cost accounts for 26%, the largest component, of the total turbine component cost. The size of tubular tower is 3–4 m (10–13 ft), with a height of 75–110 m (250–370 ft) is however size is dependent on the size of the turbine and its location. The taller the height, the more it is subjected to more wind with higher speed. Usually, the height of the tower is the

TABLE 6.1

Comparison of HAWT and VAWT [12]

Performance	Horizontal Axis Wind Turbine	Vertical Axis Wind Turbine
Generated power efficiency (%)	50–60	70
Electromagnetic interference	Yes	No
Mechanism of wind steering	Yes	No
Gearbox mechanism	Yes	No
Blade rotation space	Quite large	Quite small
Wind confrontation capability	Weak	Strong
Effect on birds	Great	Small
Ground projection effect on the human being	Dizziness	No effect
Noise (dB)	5–60	0–10
Cut in wind speed(m/s)	2.5–5	1.5–3
Failure rate system	High	Low
Operation and maintenance	Complicated	Convenient
Revolving speed	High	Low
Cable position problem	Yes	No

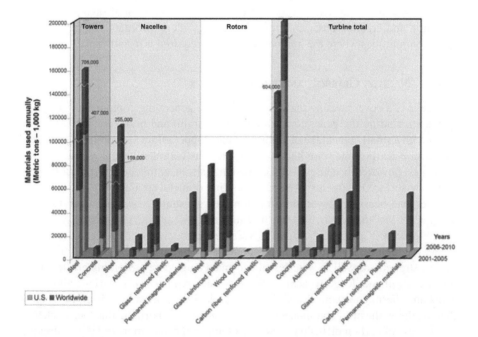

FIGURE 6.5 Wind turbine material usage trend in the US and worldwide [20].

TABLE 6.2

Distribution of Various Components of Weight and Cost of Wind Turbine [20]

Component	% of Machine Weight	% of Machine Cost
Rotor	10–14	20–30
Nacelle and machinery, less	25–40	25
Gearbox and drive train	5–15	10–15
Generator systems	2–6	5–15
Weight on top of tower	35–50	N/A
Tower	30–65	10–25

same as the diameter of the rotating circle of blades. However, to manage buoyancy force, an increase in height is associated with an increase in the diameter of the tower which correspondingly increases the cost of material as well as transportation.

The type of metal and the kind of alloy vary depending on the size and part of the wind turbine. For tower and foundation, aluminum (Al) alloy (classified aluminum alloys including the AA 3103 and 5052 because of their good cost-efficiency) or steel is widely used. For a small turbine, Al alloys are typically chosen to decrease the cost of production; while for a large turbine, higher structural strength is required for steels (structural steels such as weathering steels), which are known as atmospheric corrosion–resistant steels.

Composite materials (polymer matrix composites) have the potential to decrease the total weight of the wind turbine towers, leading to substantial savings in transportation and erection costs, making wind energy more affordable for remote and rural communities where the number of turbines required is usually small [7,13,14].

6.3.2 NACELLE, GEARBOX, AND THE ROTOR HUB

A nacelle is a cover housing that houses all of the generating components in a wind turbine, including the generator, gearbox, drive train, and brake assembly. Carbon steels with a small percentage of aluminum and copper alloys are commonly used for the nacelle gearbox and rotor hub. Carbon steels are commonly employed for nacelle bedplates; however, because of low corrosion resistance, this bedplate must be protected by surface treatments like epoxy painting or metal spraying.

The load-bearing power transmission gears usually used in wind turbines are 18CrNiMo7-6, gears. Nickel-chromium-molybdenum SAE 4820, SAE 9310, 18CrNiMo7-6, 3CrMo, 3CrMoV, and 3NiCrMo are widely used. The general requirements for high-performance gear components are a hard case providing adequate fatigue strength, as well as wear resistance and a tough core preventing brittle failure under high impact load. To achieve these surface requirements, various alloy concepts and thermomechanical and thermochemical treatments have been developed. Among these, the most common technique adopted is carburizing (the thermochemical process of carbon diffusion at elevated temperature) to improve fatigue strength and wear resistance. However, there is a potential for the replacement of carburizing with nitriding (thermochemical diffusion of nitrogen in steel) because the process temperature of nitriding is relatively lower, allowing greater flexibility during surface treatment.

Wind turbine generator converts wind energy to electrical energy. Since copper has excellent electromagnetic conductivity, a larger percentage of copper alloys is found in generators [15,16].

The rotor transfers the kinetic energy from the wind to a revolving shaft that drives a generator that produces electric power. The rotor in the generator has a magnetic field, which is created by either permanent magnets or electromagnets.

Stainless steels have high hardness, good corrosion resistance, and mechanical properties, hence they can be used for rotor hubs and blades. However, stainless steel is heavy; hence to reduce weight and decrease cost, Al alloys of series 2XXX, 3XXX, 6XXX, and 7XXX can be used [17].

6.3.3 ROTOR BLADES

Rotor blades are subjected to various structural loading effects (cyclic tension and compression loads), lightning strikes, physical impacts, and damaging surface erosion conditions while in operation. To reduce leveraged cost per unit of electricity, there is a trend to increase the size of rotor blades. However, this increase in size is associated with an increase in the cost and weight of the blade. Hence, for stability and weight reasons, composites are used.

The progress in turbine technology is associated with a continuous increase in usage of composites as the trend will be toward high strength, fatigue-resistant materials. Depending on the types of reinforcement used, the composites may be classified as glass fiber, carbon fiber, aramid fiber, or hybrid composites.

Since the fiber's stiffness and volume content determine the stiffness of composite, the maximum acceptable proportion of E glass is 65% by volume; otherwise, there will be dry areas without resin between fibers, and the fatigue strength of the composite is reduced. The advantages of glass fibers are their low-specific weight combined with high strength, good chemical, thermal and electrical properties, easy handleability, good processing qualities, and general availability at a competitive price. Another alternative to E glass is carbon fibers having much higher stiffness and lower density, allowing thinner, stiffer, and lighter blades. However, the major drawbacks are low damage tolerance, low compressive strength and ultimate strain, low negative coefficient of thermal expansion, conductive, and more expensive than E glass composite. However, carbon fiber composites are sensitive to fiber misalignment and waviness: even small misalignment leads to a strong reduction of fatigue strength [18,19].

In the case of aramid fibers besides their high specific mechanical strength, these fibers are extremely tough, damage tolerant and have a high acoustic velocity, enabling them to dissipate energy quite effectively, which explains their use in all impact resistant applications like ballistic parts. The disadvantages of this fiber are a compressive strength far lower than that of carbon, relatively poor adhesion to resins, degradation on exposure to ultraviolet (UV) light, and its tendency to humidity absorption. Hybrid reinforcements (E-glass/carbon, E-glass/aramid, etc.) represent an interesting alternative to pure glass or pure carbon reinforcements. Full replacement would lead to 80% weight savings and cost increase by 150%, while a partial (30%) replacement would lead to only a 90% cost increase and 50% weight reduction for an 8 MW turbine. In certain cases, natural fibers – for example, sisal, flax, hemp, jute, or bamboo – can be used. These natural fibers have advantages of low costs, easy availability, and environmental-friendly; however, they have low thermal stability and a tendency to absorb moisture.

The matrix material most commonly used is either thermoset, which accounts for nearly 80% of market or reinforced composite or thermoplast [18]. The advantages of thermoset are the possibility of curing at room temperature and lower viscosity, which simplifies the infusion process and increases processing speed. Thermoplastics have advantages like they can be recycled but have disadvantages of high processing temperature which may damage the fibers and higher viscosity which is problematic for manufacturing very large blades.

The latest development includes the use of nanoengineered polymers and composites for blade manufacturing. The literature demonstrates that the addition of a small amount of 0.5 wt% of nanoreinforcement (carbon nanotubes or nanoclays) in the polymer matrix of composites, fiber sizing to nano or interlaminar layers increases mechanical properties like fatigue resistance, compressive or shear strength, and fracture toughness by 30%–80% [18].

TABLE 6.3

Various Materials Used for Large and Small Turbines [20]

	Large Turbine (Small Turbine[a])							
Component/ material % by weight	Permanent magnetic material	Prestressed concrete	Steel	Al	Cu	Glass-reinforced plastic[c]	Wood epoxy[c]	Carbon filament reinforced plastic[d]
Rotor								
Hub			(95)–100	(5)				
Blades			5			95	(95)	(95)
Nacelle[b]	(17)		(65)–80	3–4	14	1–(2)		
Gear box[b]			98–(100)	(0)–2	(<1)–2			
Generator[b]			(20)–65		(30)–35			
Frame, machinery, and shell			85–(74)	9–(50)	4–(12)	3–(5)		
Tower								

Notes:

[a] Small turbines with rated power less than 100 KW listed in italics were different.

[b] Assumes nacelle is 1/3 gearbox, 1/3 generator, and 1/3 frame and machinery.

[c] Approximately half the small turbine market (measured in MW) is direct drive with no gearbox.

[d] Rotor blades are wither glass-reinforced plastic, wood epoxy, or injection molded plastic with carbon fibers.

6.4 TRIBOLOGICAL FAILURE ANALYSIS OF WIND TURBINE COMPONENTS

As mentioned in the introduction, wind energy is one of the fastest-growing renewable energy sources. However, there is an increase in operation and maintenance costs due to premature main component failure. Hence, to enhance reliability and availability, condition monitoring system (CMS) that detects turbine fault in the early stage is required. The main components are the focus of all CMS since they cause high repair costs and component downtime. Most of the failure that occurs due to bearing can be mainly classified into three types according to the kind of used sensor or monitoring object:

i. Vibration-based monitoring (used to monitor bearing faults in machine components)
ii. Oil debris-based monitoring (deterioration of lubricant) and
iii. Blade monitoring [21]

6.4.1 Vibration-Based Monitoring (Focus on Bearing Failure)

Figure 6.6 indicates that gearbox, generator, and blade constitute the maximum cost and downtime for any maintenance of wind turbine. Recently, it has been known that the around 70% of gearbox downtime and 21%–70% of generator downtime (21% on small generators ($P < 1$ MW), 70% on medium generators (1 MW $< P < 2$ MW), and 50% on large generators ($P > 2$ MW)). Figure 6.7 illustrates the downtime for gearboxes and medium generators. Hence, it is crucial to use condition-based maintenance on bearings and also to develop tools and methods that can be used to early diagnose faults.

The bearings in wind turbines can be classified in terms of rotation speed: (1) very slow-speed bearings (yaw bearings and pitch bearings), (2) slow-speed bearings (main bearings and gearbox low-speed bearings), and (3) high-speed bearings (gearbox

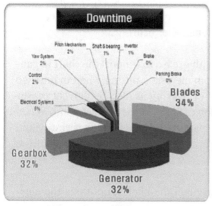 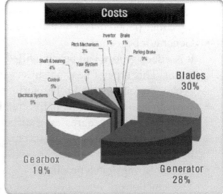

FIGURE 6.6 Failure rate of wind turbine component [21].

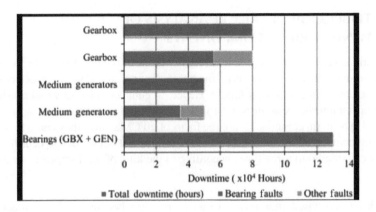

FIGURE 6.7 Aggregated downtime for gearbox and medium generators (1 MW$<P$ <2 MW) [22].

high-speed bearings, gearbox intermediate-speed bearings, and generator bearings). Their different rotation speeds and unique working mechanisms lead to different failure modes.

Various types of large-scale wind turbine bearings are as follows:

- Main bearings: (Main shaft bearing)
  The function of the main bearing is to hold hub and blades and transmit torque to gearbox. It has to work in a harsh and unstable environment. The types of main bearings are normally chosen as spherical roller bearings, cylindrical roller bearings, and tapered roller bearings (of size greater than 1 m) as they have to support heavy radial load. This bearing can be protected by frequent greasing with an adequate amount of lubricant.
- Gearbox bearings:
  Gearbox bearing in wind turbine gearbox supports shafts in different gear stages, i.e., a low-speed stage, intermediate-speed stage, and high-speed stage. The size of the bearing varies with respect to the type of load. The choice of bearing varies from cylindrical roller bearings, tapered roller bearings, four-point contact bearings, and tube roller bearings.
- Generator bearings:
  It is a deep groove ball bearing filled with grease to support the shaft in generator. This type of bearing suffers from electrical pitting wherein damage occurs on raceway and roller surfaces of bearing as electric current passes through bearing resulting in the creation of spark at the contact point. When electrical pitting progresses, a striped pattern develops on the raceway and rolling surfaces resulting in severe vibration and improper function of the bearing.
- Blade bearings/pitch bearings:
  The functions of blade bearings are to pitch the blades, optimize the electrical energy output, and stop the wind turbine for protection if the wind speeds are larger than a cut-out speed.

- Yaw bearings:
  The yaw bearings in wind turbines are used to angularly realign the nacelle into the predominant wind direction in order to optimize power input.

  Different types of bearings have different designed function and operation conditions and hence different failure modes, including multiple failure modes simultaneously. The tribological research has evolved new information and detailed knowledge describing the mechanism of failure. The data in Figure 6.8 indicates that improper lubrication is the most common reason for bearing failure and accounts for nearly 80% of breakdowns (Figure 6.8). However, depending on operating conditions and maintenance, the proportion may vary [23,24].

  Other reasons for bearing failure may include plastic deformation of bearing may be due to overloading, misalignment, or loose fit. The details are discussed below.

  Plastic deformation is very complicated, and its progress is irreversible, It is a type of indirect failure which is caused due to the worst operating conditions. The specific types of plastic deformation are explained as follows:
- Excessive load/overloading:
  Overloading of a stationary bearing by static load or shock load leads to plastic deformation at the rolling element/raceway contacts, i.e., the formation of shallow depressions or flutes on the bearing raceways in positions corresponding to the pitch of the rolling elements. In the case of large-scale wind turbine bearings, transient wind loads can easily give rise to overloading, thus leading to plastic deformations in a bearing.

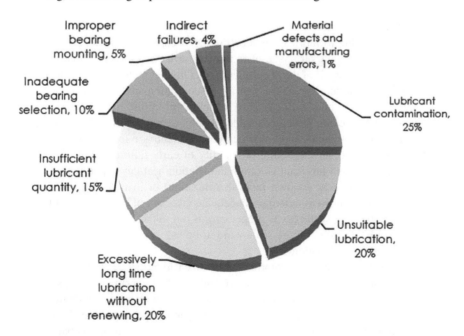

FIGURE 6.8 Common bearing failure causes in % [24].

- Misalignment:
  Load imbalance, misalignment, or improper load distribution are the main reasons for bearing failure due to improper installation, and this accounts for 5% of premature failures. A change of misalignment of 0.01/10 mm is enough to cause a huge rise in vibration and temperature in the bearing. Misalignment is a very common problem for transmission system bearings as the bearings are mounted on rubber bushings. The bushings may be deflected under the condition of large torque and extreme environment, both of which will cause bearing misalignment.
- Loose fit/incorrect seating fit:
  It occurs if the bearing ring is in relative motion to a mounting shaft or house or due to deterioration of rubber bushings in the transmission system bearing. If there is continuous relative motion it leads to vibration, fretting followed by adhesive wear, overheating, and finally leading to noise and cracking.

6.4.2 OIL DEBRIS-BASED MONITORING (DETERIORATION OF LUBRICANT)

For the continuous motion of parts, lubricants are used between contact surfaces. Further, to reduce or avoid metal-to-metal friction between rolling and sliding contact surfaces, lubricants are added. Additional functions of lubrication include heat dissipation from the bearing, removal of solid wear particles and contaminants from the rolling contact surfaces, corrosion protection, increase of the sealing effect of the bearing seals. Thus, for proper functioning and maintenance of wind bearing, lubrication is done. The following are the two most common failure modes:

- Lubricant Failure:
  The failure due to lubricant is identified as a change in grease color, and there is either a darker shade of the grease or jet black color. In addition, there will be a smell of burnt petroleum oil. Other symptoms include abnormal rise in temperature, noise, bearing discoloration, and inadequate lubricant viscosity. The identified cause for lubricant failure is the selection of incorrect lubricant. Which is unable to achieve maximum efficiency and endurance, dirty lubricants (in which contaminants are found which act as abrasive compound increasing probability of early failure) should be discarded, too much lubricant (wherein the friction heat developed within the lubricant will cause its own rapid deterioration) or inadequate (frictional heat due to under-lubrication) or inadequate viscosity of lubricant (unable to separate bearing surfaces). If proper care is not taken, this may lead to the catastrophic failure of large-scale wind turbines.
- Lubricant contamination
  Particle contamination in lubricants has been identified as a major cause of premature bearing and gear failure and can dent bearing raceways. The detrimental effect of debris contamination begins with surface indentation. Wear debris, residual particles, dirty lubricants, and water are the most common sources of contamination. Figure 6.9 shows wear debris particles embedded in the gearbox bearing roller surface.

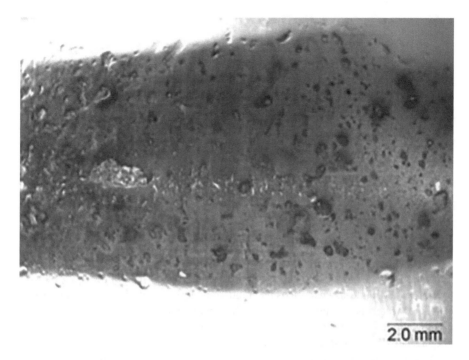

FIGURE 6.9 Embedded wear debris particles on the wind turbine gearbox bearing roller surface [2].

Lubrication quality and oil cleanliness play an important role in the deterioration rate of gearbox rolling elements. Literature indicates that poor oil cleanliness reduces the lifetime of new gearbox components by up to 50% in comparison to a well-lubricated gearbox [2,16,24,25].

6.4.3 BLADE MONITORING

As per reference [7], up to 40% of the world's energy supply will be based on renewable energy sources by the year 2030. The average capacity of wind turbines installed in 1987 was less than 50 kW, and it has increased by a factor greater than 30–1.5 MW in 2003.

Among the various important design factors, turbine blade area affects the power generated by wind turbine, and the relation is depicted by the Betz law. The power P in Watts is given by

$$P = \frac{1}{2}\alpha\rho A v^3 \tag{6.1}$$

where v is the upwind speed (i.e., the wind velocity that is incident on the turbine) in meters per second (m/s), A is the area mapped out by the turbine blades in m³, ρ is the density of air in km/m³, and $\alpha < 0.593$ is the coefficient of performance.

Figure 6.10 shows a schematic diagram of the increasing size of the blade area (i.e., rotor diameter) that has and continues to evolve in order to produce wind

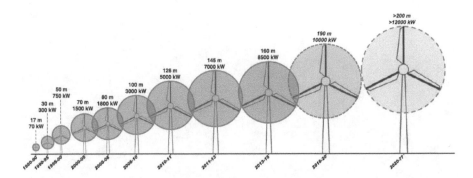

FIGURE 6.10   Blade length and rated power trend of wind turbine [26].

turbines with greater power output. However, increasing the blade size poses substantial design challenges in terms of strength, stiffness, and weight of the blade. Wind turbine blades, which are normally manufactured by continuous glass fiber-reinforced resin matrix composites, account for nearly one-fifth of the cost of a wind turbine. Since blade size as well as its manufacturing cost contributes to wind turbine efficiency in terms of total failure, downtime, and repair and/or manufacturing cost, as shown in Figure 6.6. Blade monitoring has been generating much interest recently and has been actively studied.

A significant damage form observed in operating turbine blades is caused by (abrasive) airborne particulates impacting and eroding the leading edge, especially toward the tip where velocities are higher. These airborne particles can cause significant blade erosion damage that reduces aerodynamic performance and hence, energy capture. Moreover, in some environments, insect debris and other airborne particles can accrete on the leading edges of wind turbine blades. Leading-edge blade erosion and debris accretion and contamination can dramatically reduce blade performance particularly in the high-speed rotor tip region that is crucial to optimum blade performance and energy capture. The details are discussed in the corrosion section [26–28].

6.5   CORROSION ASPECTS OF WIND TURBINES

Among the two most common types of wind turbines, offshore wind turbines are placed in locations where wind speeds are both higher and more reliable than onshore locations hence more preferred nowadays. However, offshore turbines are placed near or in the coastal areas, hence constantly exposed to marine water. In addition to humid and salty air near the offshore environments, the structures have long-term exposure to humidity with high salinity, intensive influence of UV light, wave action, and bird droppings, which decrease the corrosion resistance. The presence of dissolved oxygen, which depends on water level and temperature, increases corrosion rate by initiating cathodic reaction and increases with increase in flow velocity of air as it provides oxygen to steel structures, thereby damaging the surface film. Further, microorganisms adhere to material in water causing corrosion. Temperature, pH, and NaCl may increase corrosion (Figure 6.11).

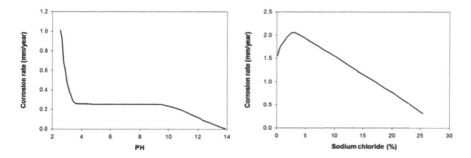

FIGURE 6.11 Relationship between corrosion rate and sodium chloride and pH [29].

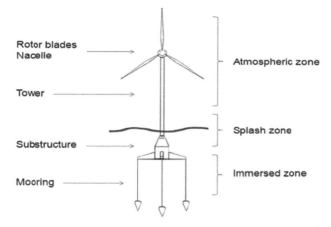

FIGURE 6.12 Various zones of offshore wind turbines [29].

The corrosion rate is dependent on the composition of environment, and since sodium chloride in the offshore environment is 3%–3.5%, severe corrosion occurs (Figure 6.11b). However, since the pH in the offshore environment is 8, the effect of pH is negligible (Figure 6.11a) [29].

As shown in Figure 6.12, there are several zones to which the tower of offshore wind energy tower is exposed including the following:

- Immersed zone (UZ): As the name indicates, this area is permanently exposed to water;
- Splash zone (SZ): In this area, there is a synergetic impact of water and atmosphere. Here water level changes due to natural or artificial effects increasing corrosion rate. This area is continuously exposed to seawater by wave and spray action resulting in high corrosion stresses.

Figure 6.12 indicates the classification of various environmental zones. Literature indicates that the highest corrosion is observed in the splash zone.

The cost due to corrosion can be bifurcated between corrosion protection (which includes steps to prevent the corrosion, maintenance cost for corrosion protection system) and corrosion damage costs (replacement of corroded parts and unplanned maintenance and repair and costs for downtime). The unplanned corrosion damage may be reduced by using corrosion-resistant materials or coatings [26,29–33].

Corrosion on offshore structures is highly dependent on site-specific factors such as water temperature, salinity, chlorinity, water depth, and current speed. The application process and the specificity of the corrosion protection system are extremely important and should be suitable for the substrate and the environment. Effective, unambiguous, feasible, and achievable specifications should be prepared by experts with a good understanding of the technology involved in protective coating systems.

6.5.1 VARIOUS FORMS OF CORROSION OCCURRING IN WIND TURBINE

Corrosion can be classified according to the appearance of the corrosion damage or the mechanism of attack as [34]:

- Uniform or general corrosion:
  There is a uniform loss of metal mass from the surface and is converted into the metal from the surface is converted to metal oxides. The extent of the metallic loss can be predicted by laboratory or field test.
- Pitting corrosion:
  There is the formation of localized pits on the surface of metal while reminder surface is unaffected, and this results in reduction in physical and mechanical properties of metal.
- Crevice corrosion:
  Similar to pitting, crevice corrosion is localized corrosion; however, there is a requirement for the presence of crevice of a very small gap, for example, in riveted plates and threaded joints.
- Filiform corrosion:
  It is observed in coated samples that are exposed to the humid atmosphere have a network of threadlike filaments.
- Galvanic corrosion:
  It is also known as bimetallic corrosion wherein metals of different compositions are in close proximity and have electrical contact.
- Erosion-corrosion:
  It includes synergetic corrosion (chemical reaction with the environment) and erosion (physical removal of the external surface) due to the movement of fluid (liquid or gas).
- Intergranular corrosion:
  Corrosion occurs near the grain boundaries due to chemical inhomogeneity in the steel, especially stainless steel.
- De-alloying:
  Generally observed in certain alloys under certain environmental conditions wherein one of the alloying elements is selectively dissolved resulting in perforation.

  Environmental-assisted cracking, including stress corrosion cracking (SCC), corrosion fatigue, and hydrogen damage: combined effect of the

TABLE 6.4

Corrosion Zones and Form of Corrosion in Offshore Wind Turbines [34]

Corrosion Zones	Main Form of Corrosion
External and internal areas of steel structure	Uniform and erosion-corrosion, SCC
Internal surfaces without control of humidity	Uniform and pitting corrosion
Internal surfaces of structural parts	Crevice, pitting and Galvanic corrosion, SCC
Splash and Tidal Zone	
External and internal areas of steel structure	Uniform, crevice, pitting Corrosion, SCC
Internal surfaces of critical structures	Uniform, crevice, pitting corrosion
Components below mean water level (MWL)	Uniform corrosion
Components below 1.0 water level of the MWL	
External surfaces in the splash zone below MWL	
Submerged Zone	
External and internal areas of steel structure	Uniform corrosion and erosion-corrosion, MIC
Internal surfaces of steel structure	Uniform, crevice and pitting Corrosion, microbiological influence corrosion (MIC)
Critical structures and components	Uniform and/or pitting corrosion MIC, SCC

chemical attack and mechanical stresses (e.g., corrosion fatigue, tensile stress with specific corrosive environments).

6.5.2 CORROSION PROTECTION

The protection of the system from corrosion is a complicated process that requires knowledge and expertise in various disciplines.

Various variables starting from composition of material, geographic and local environment, intensity of the corrosive environment, shape and size of the component to be protected, loading condition on component, and finally the life of the component required determine the corrosion protection system.

The wind turbine location environments and weather conditions have a significant impact on the operation and maintenance process, which are directly linked to energy production cost, performance, efficiency, and life of the device.

Corrosion protection may be classified into two broad systems: (1) active methods and (2) passive methods. However, in service, most protection systems use both methods. From a formal point of view, the corrosion protection methods can be classified into active or passive methods, though it is important to stress that most protection systems use both methods of protection.

6.5.2.1 Active Corrosion Protection: Cathodic Corrosion Protection (CCP)

The following methods can be considered to be active corrosion protection methods:

   i. selection of corrosion-resistant materials,
   ii. corrosion allowance,

iii. appropriate constructive design,
iv. influencing the corrosive medium,
 v. cathodic corrosion protection

Cathodic protection (CP) is a technique used to control the corrosion of a metal surface by making it the cathode of an electrochemical cell [32,35].

Further, the cathodic corrosion potential can be divided into two main classes:

(1) Galvanic Anode Cathodic Protection (GACP): This uses sacrificial anodes to protect the main structure. It is more preferred for offshore wind turbines. Zn and Al are most commonly used as galvanic anode and (2) Impressed Current Cathodic Protection (ICCP): This consists of anodes connected to a DC power source that imposes protective current. Any submerged structure irrespective of size can be protected by varying the size of anode and current requirements. In addition, the ICCP system continuously monitors the level of protection and adapts to the current required to avoid corrosion. However, it is less popular than GACP as it is more susceptible to environmental damage and mechanical damage [36].

6.5.2.2 Passive Corrosion Protection

This involves the requirement of coating for corrosion protection. Coatings improve the operational life of wind turbines. Generally, the surface of blades and towers of wind turbines are coated. Various types of coatings are polymer coatings, ceramic coatings, and metal coatings. As seen in Figure 6.13, the consumption of polymer coatings is high across the globe due to its excellent cost-to-performance ratio. Three types of coating are most commonly applied: (1) nonmetallic coatings (including organic or inorganic layers), (2) metallic coatings, or (3) the combinations of these two types of coatings on the steel surface to form a multilayer protective system (named duplex or triplex) [37].

FIGURE 6.13 Market utilization of various types of coating [38].

Both the types of turbines – offshore and onshore – need coatings to enhance the service life along with the best performance for their service life with the least maintenance. Various components of wind turbines viz blades, towers, nacelles, foundations, and equipment are coated. Among all the components, coating on blades is most important as they are continuously exposed to weather and moving in the air. Unstable harmonics is created due to pitting, which results in a decrease in turbine efficiency as well as an increase in maintenance and repair costs. The choice of the coating composition is dependent on environmental conditions, service life, required permanence, and composition of the substrate, the purpose being optimum performance. Depending on the composition, various types of coatings can be classified as:

- **Polymer coatings**
  Epoxy, polyurethane, acrylic, and fluoropolymers
  The most widely used polymer coatings are epoxy, polyurethane, acrylic, and fluoropolymer. The polymer coatings may be applied in the factory or in situ depending on equipment restrictions. The advantages of polymer coatings include high-performance properties: corrosion resistance, durability, chemical resistance, smooth finish, and toughness.

  Polymer coatings are chosen for their high-performance properties: corrosion resistance, durability, chemical resistance, smooth finish, and toughness. Fluoropolymers have the added benefit of dirt resistance and to combat biofouling at the turbine base.

  Another type of coating commonly used in powder coatings in which an organic coating is applied as a free-flowing, dry powder. Powder coatings perform better than liquid coatings and are resistant to chips, scratches, wear, and fading, The powder used for coating may be either thermoset or thermoplastic material. Because of more stability of thermoset, they are widely used for applications where high-temperature resistance service and chemical resistance are most important. However, they have the disadvantage of high maintenance cost and they cannot be repaired. The major advantage of thermoplastic is that it can be melted and remelted and easy maintenance. The most common polymers used are polyester, polyurethane, polyester-epoxy (known as hybrid), straight epoxy (fusion bonded epoxy), and acrylics.

Rubber linings
It is widely used for corrosion and abrasion resistance for protecting carbon steel equipment because of strong adhesion in the most aggressive processes of the chemical industry. For wind turbine applications, Neoprene One (chloroprene rubber) is most widely used because of its flexibility and excellent resistance to ozone, seawater, and weathering. Hence, it is used in most corrosion-prone regions of wind turbines, i.e., splash zone.

Epoxy – Aluminum oxide composite coatings
Composites are a combination of two types of materials. To increase adhesion and flexibility for increased corrosion resistance, a composite of epoxy novolac polymers and salinized micro-ceramic aluminum oxide charges are innovative coatings in application.

The latest development in polymer-based coatings includes anti-icing or icephobic surfaces that can be applied to a number of engineering applications that require the prevention of icing and easy removal of ice, such as wind turbine blades. The presence of ice on wind turbine rotors results in an imbalance of weight between turbine blades, changes in aerodynamic performance, mechanical vibrations, decrease in turbine efficiency, increase in maintenance time, and higher operating cost [38,39].

The hydrophobic surfaces can be generated by microscale or nanoscale roughness. Several important low-surface-energy polymers, for example, polytetrafluoroethylenes (PTFE) and polydimethylsiloxane (PDMS), can be used to develop these water-repellent surfaces. In reference [40], low-cost spray coating of polytetrafluoroethylene (PTFE) particles are identified and successfully deposited having micro-scale roughness and low ice adhesion strength.

- **Metallic coatings – zinc and aluminum**
  Generally, nonferrous metals, commonly zinc, aluminum, and its alloys, are used. The metallic coating protects the structure by both galvanic action and barrier. Sprayed metals are usually applied to flange connections, frames, and platform railings. However, they are also frequently applied to the whole tower sections (usually in combination with organic coating systems).

  Hot-Dip Galvanizing
  In this process, the steel structure is dipped in a molten zinc bath. The variables are immersion time, composition, size and thickness of the workpiece, and surface preparation of the workpiece. There is the presence of various sublayers in a galvanized layer which consists of varying amounts of Zn-Fe and Fe content increasing toward the substrate. This property is important for abrasive resistance compared to other coatings. The addition of Mg and Al in Zn coating improves mechanical properties as well as corrosion resistance. In the case of Al-added Zn-galvanized coating on steel, there is the formation of ductile nonbrittle of FeAlZn instead of FeZn brittle intermetallic [41].

  Thermal Spraying:
  The most applied technique for the application and repair of metallic coatings on steel structures in the offshore sector is thermal spraying on particles that are too large to be deposited in the galvanizing bath. Among several thermal spray processes most commonly used are flame spray and electric arc. The process involves propelling molten or semi-molten metal particles toward substrate by a stream of air creating layer by layer deposition till the desired thickness of the coating is achieved [37,42,43].

- **Ceramic coatings – aluminum oxide, aluminum titania, and chromium oxide**
  Ceramic coating has advantages like very high hardness, low friction, low abrasion and wear resistance, anti-galling, good corrosion resistance,

TABLE 6.5

Alternative Coating System That Can Be Used on Own Structures, Classified According to Exposure Zone [34]

Atmospheric Zone

Vinyl Systems (3–4 layers)

Zn phosphate pigmented two-pack epoxy primer (one layer)

Two-pack epoxy (two layers)

Chlorinated rubber system (3–4 layers)

Submerged Zone

The main control is CCP (Cathodic Control Protection). The use of coating systems is optional, generally EP-based coatings, and these should be compatible with CCP. When coatings are used, fewer anodes are necessary and the corrosion protection system is expected to last longer

Splash and Tidal Zone

Coatings similar to those for the atmospheric zones are used. Higher film thickness is employed

The steel thickness is increased (to act as corrosion allowance) and is coated with the same coating system as the rest of the structure

Thick rubber or neoprene coating up to 15 mm thickness

Polymeric resins or glass-flake reinforced polyester materials are often used to protect against mechanical damage

thermal stability, and high durability. Rotating machinery such as bearings needs coating to increase their lifetime while decreasing maintenance costs. The most commonly used ceramic coatings are aluminum oxide, aluminum titania, chromium oxide, etc.

Due to higher abrasion resistance (important to increase leading edge erosion resistance) than polymer coating, ceramic coating is now applied on blades of wind turbines. Although polymeric coating can provide barrier protection, it has a drawback as it is permeable to water and oxygen, resulting in poor performance in the marine environment. In addition, they have less resistance to solid particle impingement [44,45].

In addition to corrosion protection, wind turbine coatings protect two most threatening hazards:

a. Leading-edge erosion:

There is mass loss and surface damage due to erosion on wind turbine blades when exposed to raindrops.

There is a constant impact of airborne projectiles such as rain, ice salt, or sand. In addition due to the wind effect, large turbines can travel up to 250 mph as they rotate. Various variables are affecting the extent of erosion which includes the force of impact of particulate on airfoil and impact velocity which depends on wind speed and rotational speed, geometric shape, and relative velocity of airfoil and impacting particle. Hence the synergetic effect of varying angle and varying velocity is observed on the surface of the rotor. Ductile and elastic materials have low erosion resistance toward particles impingement at low angles, and brittle materials have high erosion resistance toward high impact angle as particles are unable to transfer a high amount of kinetic energy to the surface. Hence characteristic of coating behavior plays important role in its life durability and total blade failure [38–40].

b. **High maintenance costs:**

There is an uninterrupted observation of coating while the operation of turbine for the life span of 15 years. A reduction of about 20% in turbine energy output is due to rain and sand. In addition, the location and position of offshore wind turbines result in a very high expense of maintenance. The best wind power coatings aim to be maintenance-free for their lifetime [41].

6.6 CORROSION TESTING

There are two main classifications of testing and assessment of corrosion protection systems [46].

6.6.1 FIELD TEST

It consists of long-term exposure tests in a real corrosive environment. There will be three galleries (underwater zone, intermediate zone, and splash zone). All specimens are to be tested for the long term. Functional maintenance of the impressed current system (cathodic protection) is part of the site test procedure.

6.6.2 LABORATORY TESTS UNDER DEFINED ARTIFICIAL STRESS CONDITIONS

The laboratory tests were subdivided into aging tests according to ISO 20340, cathodic disbonding tests, and tests based on electrochemical impedance spectroscopy (EIS) [46] (Figure 6.14).

The aging resistance test procedure, which consists of a 7-day cycle, is exposed to three loading mechanisms and passed for 25 times thus in total 4200 hours. Adhesion strength, coating deterioration, and delamination are considered as assessment criteria for the cathodic disbonding test.

The proper information is not provided by cathodic disbonding about the mechanism of corrosion in progress or paint degradation process and are accept or reject criteria test and give a comparative evaluation of different paint systems.

A promising method for gathering degradation and corrosion information on coating for wind energy towers is EIS. Usually, EIS is carried out in 3% NaCl solution for many days (>60 days). The measurements were performed according to three-electrode methods. The barrier resistance is determined, and the extent of corrosion can be evaluated visually [46].

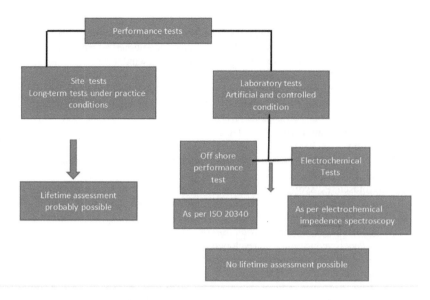

FIGURE 6.14 Summary of performance test [46].

6.7   CONCLUSION AND FUTURE WORK

Wind turbine technology offers a cost-effective alternate renewal energy source and has experienced immense growth with respect to both the turbine size and market share. Wind turbines are mostly installed in a remote area (onshore) or in/near the seashore (offshore) condition. Hence, they are working in a complex natural environment that includes stormy winds, sand particles, and salty fog. Because of their location, the operation and maintenance of wind turbines are challenging in terms of time and cost. Wear and corrosion of major components of wind turbines are reasons for breakdown. The attempt has been made to discuss the wind turbine starting from major components of wind turbine and their material properties followed by major cause of failure and possible solutions.

The research in wind turbines is directed in multiple directions. There is advance in wind turbine design and manufacturing which includes development and innovation in manufacturing, composite material development, and modeling of blades with the purpose of more reliable, less expensive, and efficient turbines. In addition, in the case of the offshore environment, towers and foundations that support the turbine are a critical part, and their design and maintenance present unique challenges to structural engineers. Another area of research includes characterizing the flow of wind as it is dependent on climatic changes as well as area of Structural Health Monitoring (SHM), Sensing, Diagnostics, and Testing [47].

REFERENCES

1. Du, Y., Zhou, S., Jing, X., Peng, Y., Wu, H., Kwok, N. Damage detection techniques for wind turbine blades: A review. *Mechanical Systems and Signal Processing* 141 (2019) 106445.

2. Liu, Z., Zhang, L. A review of failure modes, condition monitoring and fault diagnosis methods for large-scale wind turbine bearings. *Measurement* 149 (2020) 107002.

3. Lee, M. H., Shiah, Y. C., Bai, C. J. Experiments and numerical simulations of the rotor-blade performance for a small-scale horizontal axis wind turbine. *Journal of Wind Eng ineering and Industrial Aerodynamics.* 149 (2016) 17–29

4. Hand, B., Cashman, A. A review on the historical development of the lift-type vertical axis wind turbine: From onshore to offshore floating application. *Sustainable Energy Technologies and Assessments* 38 (2020) 1006462.

5. https://www.saurenergy.com/solar-energy-news/global-wind-energy-market-to-reach-124-6-bn-by-2030.

6. http://www.ewea.org/fileadmin/files/library/publications/reports/Economics_of_Wind_Energy.pdf.

7. Enevoldsen, P., Xydis, G. Review examining the trends of 35 years growth of key wind turbine components *Energy for Sustainable Development* 50 (June 2019) 18–26

8. Bhardwaj, U., Teixeira, A. P., Guedes Soares, C. Reliability prediction of an offshore wind turbine gearbox. *Renewable Energy* 141 (October 2019) 693–70.

9. https://energyeducation.ca/encyclopedia/Wind_turbine.

10. Saad, M. M. M., Asmuin, N. Comparison of horizontal axis wind turbines and vertical axis wind turbines. *IOSR Journal of Engineering (IOSRJEN)* 04(08) (August 2014) 27–30.

11. https://www.windpowerengineering.com/vertical-axis-wind-turbines-vs-horizontal-axis-wind-turbines/.

12. www.windturbinestarj/hawt-vs-vawt.html.

13. O'Leary, K., Pakrashia, V., Kelliher, D. Optimization of composite material tower for offshore wind turbine structures. *Renewable Energy* 140 (September 2019) 928–942.

14. Polyzois, D. J., Raftoyiannis, I. G., Ungkurapinan, N. Static and dynamic characteristics of multi-cell jointed GFRP wind turbine towers. *Composite Structures* 90(1) (2009) 34–42.

15. Tobie, T., Hippenstiel, F., Mohrbacher, H. Optimizing gear performance by alloy modification of carburizing steels. *Metals* 7 (2017) 415.

16. Zhou, J., Roshanmanesh, S., Hayati, F., Papaelias, M. Improving the reliability of industrial multi-MW wind turbines. *Insight - Non-Destructive Testing and Condition Monitoring* 59(4):189–195

17. Islam, M. R., Guo, Y., Zhu, J. A review of offshore wind turbine nacelle: Technical challenges, and research and developmental trends. *Renewable and Sustainable Energy Reviews* 33 (May 2014) 161–176.

18. Haberkern H. Tailor-made reinforcements. *Reinforced Plastics and Composites* 50 (2006) 28–33.

19. Mishnaevsky, Jr., L., Branner, K., Petersen, H. N., Beauson, J., McGugan, M., Sørensen, B. F. Materials for wind turbine blades: An overview. *Materials (Basel)* 10(11) (November 2017) 1285.

20. Ancona, D., McVeigh, J. Wind Turbine - Materials and Manufacturing Fact Sheet Prepared for the Office of Industrial Technologies, US Department of Energy By Princeton Energy Resources International, LLC.

21. Lee, J. K., Park, J. Y., Oh, K. Y., Ju, S. H., Lee, J. S. Transformation algorithm of wind turbine blade moment signals for blade condition monitoring. *Renewable Energy* 79 (2015) 209–218.

22. de Azevedo, H. D. M., Araújo, A. M., Bouchonneau, N. A review of wind turbine bearing condition monitoring: State of the art and challenges. *Renewable and Sustainable Energy Reviews* 56 (April 2016) 368–379.

23. Yagi, S., Ninoyu, N. Technical trends in wind turbine bearing. *TechRev* 76 (2008) 113–120.

24. https://www.bearing-news.com/the-most-common-causes-of-bearing-failure-and-the-importance-of-bearing-lubrication/.
25. Paskaa, J., Sałeka, M., Surma, T. Current status and perspectives of renewable energy sources in Poland. *Renewable and Sustainable Energy Reviews* 13(1) (January 2009) 142–154.
26. Bartolomé, L., Teuwen, J. Prospective challenges in the experimentation of the rain erosion on the leading edge of wind turbine blades. *Wind Energy* 22 (2019) 140–151.
27. https://www.machinerylubrication.com/.
28. Ai, X. Effect of debris contamination on the fatigue life of roller bearings. *Proceedings of the Institution of Mechanical Engineers, Part J: Journal of Engineering Tribology*, 215(6) (2001) 563–575.
29. Choi, S. J., Park, M. H., Kim, I. T. Corrosion-Protection Design for Floating-Type Offshore Wind Turbines. *The 2012 World Congress on Advances in Civil, Environmental, and Materials Research (ACEM' 12)*, Seoul, Korea, August 26–30, 2012.
30. Willett, H. G. Characterisation of composites for wind turbine blades. *Reinforced Plastics* 56(5) (September–October 2012) 34–36.
31. Sareen, A., Sapre, C. A., Selig, M. S. Effects of leading edge erosion on wind turbine blade performance. *Wind Energy* 17 (2014) 1531–1542.
32. Momber, A.W., Plagemann, P., Stenzel, V., Schneider, M. Investigating corrosion protection of offshore wind turbines. *Journal of Protective Coatings and Linings* 25(4) (2008) 30–43.
33. Momber, A. Corrosion and corrosion protection of support structures for offshore wind energy devices (OWEA). *Materials and Corrosion* 62(5) (2010) 391–404.
34. Source: NeSSIE Project – cofounded by the European Maritime and Fisheries Fund (EMFF) – www.nessieproject.com State of the Art Study on Materials and Solutions against Corrosion in Offshore Structures North Sea Solutions for Innovation in Corrosion for Energy Third edition, revised and updated - February 2019 © NeSSIE - North Sea Solutions for Innovation in Corrosion for Energy – 2019.
35. Price, S. J., Figueira, R. B. Review corrosion protection systems and fatigue corrosion in offshore wind structures: Current status and future perspectives *Coatings* 7 (2017) 25.
36. https://en.wikipedia.org/wiki/Cathodic_protection.
37. https://cathwell.com/industries/wind-turbines/9527-2/#:.
38. https://www.transparencymarketresearch.com/wind-power-coating-market.html.
39. Pugh, K., Rasool, G., Stack, M. M. Some thoughts on mapping tribological issues of wind turbine blades due to effects of onshore and offshore raindrop erosion. *Journal of Bio- and Tribo-Corrosion* 4 (2018) 50.
40. Liang, F., Gou, J., Kapat, J., Gu, H., Song, G. Multifunctional nanocomposite coating for wind turbine blades. *International Journal of Smart and Nano Materials* 2(3) (2011) 120–133.
41. Valaker, E. A., Armada, S., Wilson, S. Droplet erosion protection coatings for offshore wind turbine blades. *Energy Procedia* 80 (2015) 263–275. 12th Deep Sea Offshore Wind R&D Conference, EERA DeepWind'2015.
42. https://knowledge.ulprospector.com/6931/pc-the-current-spin-on-wind-turbine-coatings/.
43. Qin, C. C., Mulroney, A. T., Gupta, M. C. Anti-icing epoxy resin surface modified by spray coating of PTFE Teflon particles for wind turbine blades. *Materials Today Communication* 22 (March 2020) 100770.
44. Syrek-Gerstenkorn, B., Paul, S., Davenport, A. J.. Sacrificial thermally sprayed aluminium coatings for marine environments: A review. *Coatings* 10(3) (2020) 267.
45. Musabikha, S., Utama, I. K. A. P., Mukhtasor. State of the Art in Protection of Erosion-Corrosion on Vertical Axis Tidal Current Turbine. *AIP Conference Proceedings* 1964, 020047 (2018).

46. Momber, A. W., Plagemann, P., Stenzel, V., Schneider, M. Investigating corrosion protection of offshore wind towers. *Journal of Protective Coatings and Linings* 25(4) (2008) 30.
47. Willis, D.J., Niezrecki, C., Kuchma, D., Hines, E., Arwade, S., Barthelmie, R.J., DiPaola, M., Drane, P. J., Hansen, C.J., Inalpolat, M., Mack, J.H., Myers, A.T., Rotea, M. Wind energy research: State-of-the-art and future research directions. *Renewable Energy* 125 (2018) 133–154.

7 Surface Texturing Practices to Improve the Wear Behavior of Cutting Tools for Machining of Super Alloys

Jibin T. Philip

National Institute of Technology Mizoram

Amal Jyothi College of Engineering

Cherian Paul and Tijo D.

Saintgits College of Engineering

CONTENTS

7.1 INTRODUCTION

With the development in aviation engine technology, superalloys are majorly used material in the new and modern engines. Utilization of these heat resistant superalloys in the manufacture of engine and its related parts allows to increase the engine temperature by 10°C (per year), since 1950s. The rise in engine temperature leads to the decrease in fuel consumption and engine efficiency [1]. Besides, these alloys are also used for applications demanding/requiring high strength to weight ratio,

DOI: 10.1201/9781003097082-7

high fatigue resistance, and oxidation resistance such as nuclear, gas turbine, rocket, biomedical, and chemical vessels. Superalloys are majorly divided into four categories, viz. (1) nickel-based; (2) iron-based; (3) cobalt-based; and (4) titanium-based [2]. Among them, nickel and titanium-based alloys are extensively used where high strength is required at elevated temperatures. For example, Inconel 718 is the widely used nickel-based alloy and exhibits a tensile strength (approx.) of 1375 N/mm^2 at room temperature and a low decrement of 20% (the value around 1150 N/mm^2) shown at elevated temperatures up to 650°C [3]. In contrast to Inconel 718 (nickel-based alloy), Ti6Al4V (titanium-based alloy) exhibits better strength up to 600°C. Moreover, during machining, these alloys exhibited high strength even at elevated cutting temperatures. Hence, they are known as difficult-to-cut materials. During machining of these alloys, due to severe plastic deformation, high cutting temperature is developed at the interface zones of tool and chip. It results in a drastic increment in tool wear leading to substantial shortening of tool life. To reduce the cutting temperature, the researchers have focused on the modification of the surface of the cutting tools. Due to friction between the two mating parts, the development of high temperature was reported at the interface regions. Hence, the researchers are primarily focused on the surface modification of the tool to reduce friction between the tool and chip. Surface texturing is the novel surface modification method, in which different kinds of textured patterns are provided on the tool rake and flank faces for reducing friction, cutting temperature, and reducing tool wear. Mainly, three mechanisms are responsible for improving the machining performance of Inconel 718 and Ti6Al4V with textured tools. They are: (1) decrease the tool-chip contact length: by providing the surface textures with the micro and nano sizes on the different faces of the cutting tool can decrease the tool-chip contact length (L_r). Then, the actual contact length (L_f) can be calculated as shown in Equation (7.1).

$$L_f = L_r - n \cdot w_g \qquad (7.1)$$

where (L_f) is the actual tool-chip length, L_r is the nominal tool-chip contact length, n is the number of grooves located at the contact region, and w_g is the width of each groove. If the nominal tool-chip contact length L_r is 1.0 mm, and there are three grooves with a width of 50 μm located in the chip flow direction at the contact zone. According to Equation 7.1, the actual contact length (L_f) will be decreased from 1.0 to 0.85 mm, and the cutting forces will be decreased by 10%. This is presumably due to a decrease in the tool-chip contact length, thereby frictional force and the associated cutting temperatures also decrease. (2) Wear debris entrapment: while machining Inconel 718 and Ti6Al4V, different types of wear mechanisms such as abrasion, adhesion, etc., are responsible for the generation/formation of wear debris at the interface surfaces. Surface texturing can help prevent third body abrasion by capturing wear debris formed at the tool-chip interface region. During machining, the textures will entrap the wear debris in dry and lubricated conditions equally. (3) Textured patterns act like fluid reservoir: the generation of textures on the flank and rake surfaces can be expected to act as a fluid reservoir that promotes the supply of the cutting fluid to critical temperature developed regions of the cutting zones and worn area of the tool.

Distinct methods are available for the generation of textures on the cutting tool (rake and flank faces). Among them, the laser surface texturing (LST) method is frequently used by researchers. In LST, a high-energy pulse is applied, and removal of material takes place by laser ablation, i.e., by rapid melting and vaporization. LST has various advantages, such as it provides proper surface topography and less contamination of the substrate surface in comparison to other methods. During machining of the superalloys (Inconel 718/Ti6Al4V), due to their high-temperature properties, superior strength is attained while machining, and these materials contain carbide particles which are responsible for severe tool wear. These are some of the major problems leading to the shortening of tool life. Hence, to improve the tribological response and machining performance of the textured tools during machining of titanium alloy (Ti6Al4V) and nickel alloy (Inconel 718), the textured cutting tools should possess: (1) good hot hardness; (2) high wear resistance; (3) high toughness and strength; and (4) better thermal shock properties at high temperatures. Moreover, adequate chemical stability is also essential. For economic and efficient machining of nickel and titanium alloys, it requires a good understanding of the cutting conditions, cutting tool materials, and the proper geometry and functionality of the machined component.

In general, HSS cutting tool material is used for machining of all conventional work materials, except superalloys. Since the thermal softening point (temperature value) is very low for HSS, i.e., around 600°C. Hence, during machining of high melting temperature alloys, HSS cutting tool material cannot exist. Therefore, carbide tools are still used for the industrial general-purpose machining applications associated with nickel and titanium-based alloys. It is because the thermal point temperature of the carbide tools is 1100°C (approx.). However, with the drastic increase in demand for better surface quality and fast material removal, high-speed machining was established. Nonetheless, the use of cemented carbide cutting tools has become more problematic. Many new developments have been made in tool surface coatings since the 1970s to significantly increase the hardness, modulus, and toughness of the tool, and thereby they will be able to give better performance in friction, wear, and lubrication applications. However, the application of such tools can only be at a lower order of cutting speeds, the limitation being due to cutting temperature.

Ceramic tools are employed to improve and achieve high-speed machining during the processing of nickel and titanium alloys. The most commonly used ceramic tools are (1) SiC whisker-reinforced alumina (Al_2O_3) ceramic, (2) silicon nitride (Si_3N_4) ceramic, and (3) TiC added alumina (Al_2O_3–TiC) ceramic. The major limitations of ceramic cutting tools are their low fracture toughness and their low thermal conductivity. The low toughness and excessive tool wear rate are the biggest problems while machining nickel and titanium alloys. Ultra-hard cutting tool materials such as cubic boron nitride (CBN) and diamond inserts have been noticed to generate good machined surface characteristics, with a considerable increase in tool wear resistance and tool life associated with their higher thermal conductivity and hot hardness. CBN and diamond have the highest thermal conductivity in contrast with ceramic cutting tools. CBN and diamond-cutting tool materials were able to remove heat efficiently from the cutting edge, and also compared to ceramic materials, these cutting materials exhibited the highest wear resistance and improved the friction coefficient. Hence, these super-hard materials are highly suited for the machining of nickel and titanium alloys. The performance of the CBN-cutting tool mainly depends on the CBN content and binder of the inserts. The predominant wear mechanisms were found to be diffusion and chipping of the cutting edge while machining of nickel alloys. Sugihara et al. [4] carried out a machining trial with developed textured inserts on Inconel 718. They have observed that by providing the texture on the flank face, the chipping of the cutting edge was reduced compared to the conventional CBN tool.

7.2 SURFACE TEXTURING

Engineering materials, including metals, composites, and ceramics, are proven to be highly dependent on the machining process for the manufacturing of various part geometries. However, a major concern is that tool wear affects the quality, productivity, and cost of the components that are being machined. The rake and flank faces of the cutting tool are subjected to wear, and the mechanism is of different types, including fatigue, abrasion, diffusion, and adhesive wear. Many other modes are also reported by other researchers [5,6]. Cutting forces, vibrations, and heat generation that are dependent on the friction at the tool-chip/work interfaces are also to be considered since they have an effect on tool wear. Many researchers have conducted investigations into reducing the wear and friction of the cutting tools during machining, and they include improved cutting tool designs [7], development and applications of nano-lubricants [8], and optimization of the various process parameters [9].

The surface modification process carried out on the cutting tools is of great interest in the current chapter. There are three different methods by which surface modification can be carried out on cutting tools, viz. (1) traditional methods like surface coating; (2) surface melting/treatment with the help of LASER/e-beam sources; and (3) surface texturing. Pondering these aforementioned processes, deposition of various elements/materials in a single or multilayered manner with a thickness at the micron level is carried out on the surface coating. Usually, the elements that are coated exhibit superior hardness, good thermal stability, and improved tribological properties [10]. Second, radiant energy sources such as LASER/e-beam are used to modify the surface of the substrate by remelting or alloying. This results in variations in the microstructure of the modified layer, resulting in a fine grain structure owing to the rapid solidification, thereby leading the alloy to exhibit superior mechanical and tribological properties. Furthermore, in comparison, recently, surface texturing has emerged as a new surface modification technique wherein surface textures are manufactured at micro/nanoscale on various cutting tools using various techniques. The design of these textures should be in such a way that it will facilitate the lubricant availability at the tool/chip interface, reduce wear rate, and aid in achieving a minimum contact area between the chip and the tool.

Surface texture, the characteristic of a surface, is also termed as the surface topography, which constitutes all the minute variations and incorrect tendency that a surface process has from the true plane that is perfectly flat upon visual examination. However, surface roughness is defined as the roughness or surface finish that comprises finely spaced microsurface irregularities that can result from various machining operations. The randomness of the inherent surface roughness can be controlled using surface texturing. It has to be noted that the lubricants serve the purpose of filling the valleys existent on the rough surfaces, which are encountered. Although lubrication is facilitated, it will not be in a controlled manner. The texture that is created artificially will become a part of the overall surface structure, if small, and will be considered in the geometry as they increase in size. Further, the improved tribological properties expected upon surface texturing can be attributed to the facilitation of lubricant entrapment and the possibility for the formation of micro wedges.

In the past, uncoated and cemented WC inserts were ground to a micro-depth on the rake surface with the aid of a diamond grinding wheel [6,11]. The cutting parameters, viz. cutting speed and feed rate, were kept constant as the different grain sizes of the used WC insert were used for the investigation. Predesigned textured cutting tools were reported to be used in the machining of steel for enhancing wear resistance [12]. Further, a newly developed cutting tool was also proposed consisting of grooves with micro-stripes, which are introduced on the rake surface [13]. The investigation showed that the insert textured with microgrooves parallel to the cutting edge helped with the resulting enhanced wear resistance and, additionally, effective lubricity of the insert surface during the milling operation with steel materials. The micro-periodic grooves designed parallel to the cutting edge of the insert effectively performed the purpose of preserving cutting fluid and minimizing the actual tool and chip contact area. Recently, LASER technology has found its application in preparing three types of textured dimples (circular and rectangular patterns, and parallel grooves) on the rake surface of a cemented carbide insert. The investigation was on the dry cutting of aluminum alloy. Textured inserts exhibited good cutting performance at low cutting speeds, and those with rectangular dimples proved to be the best choice [14]. It was also reported that the frictional behavior at the tool/chip-workpiece interface was reduced while using micro or nano fine-scaled patterns and structures on the rake surface of the cutting tool [15].

Various methods have been developed to fabricate textured surfaces [16]. The advantages of each of the methods are based on the substrate or parent material, expected precision level, production cost, and quality. A few include ultraviolet photolithography, etching with LASER source, electrical discharge material removal, etc. It was also reported that the application of various methods such as micro indentation, blasting, abrasive jet machining (AJM), etc. Given the ease of operation as well as better cost-effectiveness, laser surface texturing (LST) is considered the most effective method of its kind. The texturing methodologies will be explained in detail in the progressive sessions. To discuss, the automotive industry has been keen on developing methods that can modify the substrate surface, thereby achieving superior physical as well as mechanical properties with operational performance. A cylinder liner textured in a cross-hatched pattern is expected to retain and supply the lubricant in the contact area of the piston ring [16]. Figure 7.1 shows the cross-hatched pattern textured on the steel cylinder liner.

The texturing process utilizes some distinctive features to incorporate those that normally consist of different shapes, including dimples, troughs, pores or small pockets, and grooves. A few are shown in Figures 7.2–7.4. The texturing arrangements are

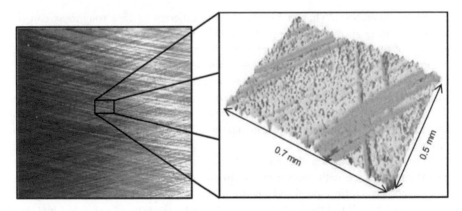

FIGURE 7.1 Cross-hatch pattern textured in steel cylinder liner [15].

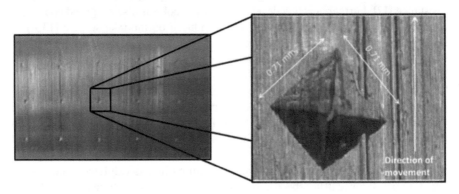

FIGURE 7.2 Dimples with pyramid-shaped textures developed using micro-indenter on a shaft (Journal bearing) [15].

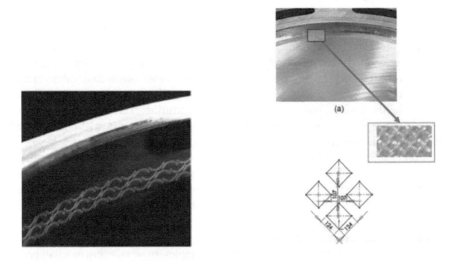

FIGURE 7.3 Laser surface texturing on cylinder liner [15].

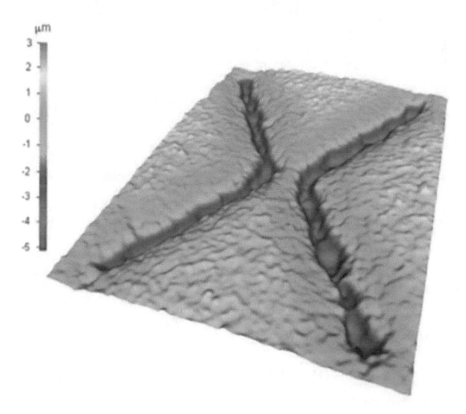

FIGURE 7.4 Laser surface texturing in chevron shape on steel plate [15].

usually hexagonal, square, or random, as shown in Figure 7.5. Figure 7.6 shows the rake face of a LASER micro-machined Chevron-shaped uncoated WC cutting tools, in an attempt to machine Ti-6Al-4V alloy [6]. The laser textured uncoated WC tool is as shown in Figure 7.6b. The dimensions of the holes textured at the micro-level

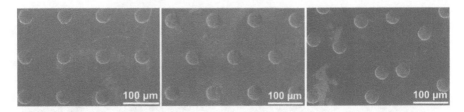

FIGURE 7.5   Square, hexagonal, and random arrangement in laser surface texturing [16].

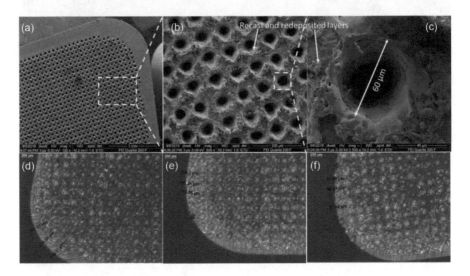

FIGURE 7.6   Laser-textured rake face of cutting tools at (a) 100X, (b) texture dimensions, and (c) laser-induced layers around micro-holes and textured tools at varying distances from cutting edges (d–f) [6].

can be seen with high magnification. The recast layer, which is the sideway deposit formed of the material that is ablated because of the laser surface interaction is illustrated in Figure 7.6c. The images of the tools that are subjected to different textures are shown in Figure 7.6d–f. It can be observed that for different textures the distance from the cutting edge is different.

7.3   MACHINING OF SUPER ALLOYS USING TEXTURED TOOLS

7.3.1   INCONEL 718 AND TI6AL4V

The enhancement of tribological and machining performance of Inconel 718 and Ti6Al4V using textured cutting tool depends on the geometry of the texture patterns such as micro-holes or micro dimples of square, elliptical, triangular, wavy grooves, groove pattern, micro pits, chevron, etc. Rathod et al. [17] studied the tribological and machining performance of circular, square, and linear textured cutting tools while machining Ti6Al4V. Among all the textured tools, square-textured inserts have shown lower cutting forces and cutting temperatures and aid in improving the tool life. Observably, it also minimizes the adhesion of the work material with the tool. Moreover, a significant reduction in heat generation and cutting forces is also possible. It has resulted (majorly) due to the reduction in contact length between the chip and tool. The textured patterns act as a reservoir of solid lubricant (WS_2). It also depends on the dimensions of the textures developed, such as texture width, pitch, depth, edge distance, and diameter. Ma et al. [18] investigated the effect of micro-hole-textured parameters, such as microhole diameter, micro-hole position, micro-hole diameter to depth, and cutting edge, through FEM analysis during machining of the Ti6Al4V. The authors noticed that by varying the micro-hole parameters, the cutting forces varied. A minimum of three cutting forces was observed at the optimal values of micro-hole diameter, edge distance, and diameter to depth ratio as considered at the values of 60–100 μm, 20–100 μm, and 10–14 μm, respectively. The micro-hole position does not influence the three cutting forces (F_x, F_y, F_z). Texture orientation with the cutting edge (of the tool) is also an important parameter that affects the machining performance of the textured tool. The orientation of the textured pattern can be considered as parallel, perpendicular, and crossed to the cutting edge. The researchers found that perpendicular-oriented textured grooves exhibited a lower coefficient of friction, wear rate, and cutting temperatures in contrast to other textured orientations. Arulkirubakaran et al. [19] investigated the effect of orientation of the textured patterns, viz. conventional, parallel, perpendicular, and cross texture (to the cutting edge), while machining of titanium alloy (Ti6Al4V). The authors have found that the perpendicular textured pattern demonstrates lower contact area, coefficient of friction, and cutting temperatures, and the perpendicular textured insert has shown a lower wear rate even at high-speed machining. It is the resultant effect of better micro-pool lubrication of MoS_2 (solid lubricant) at the interface of tool and chip with minimum contact length.

In the context of machining of Inconel 718 and Ti6Al4V, the need remains to develop a wear rate theory or model for predicting the tool wear of textured cutting tools. The researchers are using the textured dimensions determined using a trial-and-error method, so it is essential to develop a model or theory to determine the optimum dimensions, orientation, and shape of the textures. Nickel and titanium-based alloys are associated with severe plastic deformation and work hardening. Hence, high cutting temperatures are developed at the interface zone of the tool and chip, so it is essential to optimize the friction at the mating surfaces of tool and chip. Besides, the hot machining process characteristics can be improved with the implementation/utilization of textured tools. Texturing can also aid in improving the tribological characteristics of the various categories of tools during the machining of superalloys. Inconel 718 and Ti6Al4V are high heat-resistant nickel and titanium-based superalloys. Processing of such alloys can lead to the development of abrasive particles, which can result in severe tool wear. Moreover, the plastic deformability and high strain rate sensitivity of these alloys favor the tendency for work hardening; hence high cutting temperatures are developed at the interface regions of tool and chip. To eradicate the friction and its corresponding temperatures, the researchers have developed a novel surface modification technique called surface texturing. In this technique, different kinds of orientation of the textures and shapes are provided on the rake and flank face of the tool to decrease the friction coefficient and tool wear and to improve the tool life. During machining of Inconel 718 and Ti6Al4V, the machining and tribological performance improved due to decrement of the tool and chip contact length, and entrapment of the wear debris and textures provided the micro-pool lubrication at the interface zone of the tool and chip. Most researchers have considered the optimal dimensions of the textures while machining these alloys based on the trial-and-error method. Hence, it is required to develop a friction and tool wear model for better improved tribological performance through the utilization of textured tools.

7.4 LASER SURFACE TEXTURING (LST)

LST is a surface modification technique that improves the tribological performance of materials. This technique is useful for altering the properties of the metallic surface without unduly changing the essential properties of the base material [20]. Surface modification by LST offers several advantages, such as short processing time, controllability, and environmentally friendly, and can fabricate different textures on mechanical components [21,22]. Considering the advantages, the LST process has been adopted to enhance the mechanical performance of various engineering components [23,24]. In recent years, using the LST process, different patterns of surface textures have been developed on mechanical components to improve their surface properties, which are versatile to wear owing to their poor tribological properties. By controlling the laser power, the morphology of the textured surface can be varied; consequently, the required mechanical properties can be achieved.

Through the continuous progress in materials in the manufacturing sector, most of the advanced materials possess excellent hardness and wear resistance. On the contrary, the machining of advanced materials turns into a demanding task owing to hasty tool wear, which further reduces the tool life. Traditionally, cutting fluid was used to minimize tool wear during the machining processes. Cutting fluids reduce the friction between tool and workpiece and minimize the temperature during machining, which further improves the life of tool. However, previous works were detailed that most of the cutting fluid consists of chemicals that are harmful to human beings. Further, recycling of cutting fluids is not economical and disposal is arduous [25]. Considering the negative impact of cutting fluids on human beings and environment, researchers have started working toward the minimization of cutting fluids while machining.

Surface texturing is an emerging technique that can be adopted to enhance the tribological performance of materials and, thereby, their life. Over the years, various researchers adopted the LST technique to fabricate textures on the flank and rake face of the cutting tool to reduce wear and friction during the contact between tool and material to be machined. Lei et al. [26] developed micro-holes on the rake surface of tungsten carbide (WC) insert close to the cutting edge by femtosecond laser to machine mild steel. They analyzed the performance of micro-hole lubricated cutting tool by assessing the performance in terms of cutting force, tool-chip contact, and morphology of the chips formed. Results revealed that tools with micro-holes bring only the slightest amount of cutting fluid into the interface between tool and chip, which further reduced the coefficient of friction and minimized the energy loss owing to the friction as compared to the traditional machining process. In the last decade or so, many researchers have successfully employed laser surface-textured cutting tools for the machining of Inconel 718 and Ti-6Al-4V alloys.

Mishra et al. [27] used laser-textured tools coated with hard ceramics for the machining of Ti-6Al-4V alloy in a dry cutting environment. They evaluated cutting force, flank wear progress, and rake surface analysis after machining. It was revealed that machining with a laser textured and coated cutting tool resulted in reduced cutting force, thrust force, and flank wear. Ze et al. [28] performed machining of Ti-6Al-4V alloy using laser-textured cemented carbide (WC/Co) inserts filled with MoS_2 lubricants in the micro-holes. They performed dry machining using textured and conventional tools. It was found that the cutting force and temperatures in dry machining of Ti-6Al-4V alloy using self-lubricating textured tools were minimized significantly as compared to traditional tools under the same machining conditions. Sasi et al. [29] analyzed the performance of laser surface-textured HSS cutting tool while machining of Al7075-T6 aerospace alloy. It was revealed that laser surface texturing can be used to fabricate micro-dimples on HSS and which can enhance the mechanical performance of cutting tool under dry conditioning. Further, it was found that by using textured cutting tools, cutting force can be reduced under low- and high-speed conditions, but it is found to be more effective at high-speed conditions.

Among the known materials on the earth, cubic boron nitride (CBN) is the second hardest behind diamond, and it has received substantial attention as a material for cutting tools. However, machining of superalloys such as Inconel 718 becomes difficult even with a tool made of CBN. Sugihara et al. [4] reported that the CBN tool exhibited crater and flank wear owing to the diffusion at high temperatures and adhesion of chips on the flank face of the tool resulted in the chipping of the cutting edge. The same authors performed the machining of Inconel 718 using the CBN tool with microgrooves orthogonal cutting edge produced with a femtosecond laser, and the results revealed that the tool wear was significantly reduced as compared to machining with conventional cutting tools. The above-reported works clearly show that laser surface texturing on the flank face of the cutting tool is a promising technique to enhance the life of the cutting tool while machining hard materials like Inconel 718 and Ti-6Al-4V alloy.

7.4.1 INFLUENTIAL FACTORS

In the LST process, the selection of process parameters is very important, since it has a significant impact on the morphology and tribological properties of the modified surface. In addition to the selection of laser types, parameters such as spot size, laser power, scan speed, energy input, and feed rate can control the quality of textured surfaces. Therefore, a study to understand the optimum conditions for accomplishing the required mechanical performance of textured surfaces makes sense. Laser power intensity is a significant process parameter in the LST process which determines the morphology and mechanical performance of the textured surface. Mao et al. [30] reported that power intensity regulates the shockwave pressure during LST processing, which further determines plastic deformation. Thus, the morphology and mechanical properties of the treated material are influenced by power intensity.

Several researchers made an effort to study the impact of laser power intensity on the morphology and mechanical performance of laser-textured workpiece surfaces. Sánchez et al. [31] revealed that the micro-hardness of treated aluminum alloy was enhanced up to 1.2 times than the untreated sample when power intensity increased from 900 to 5000 pul/cm^2. Further, the improvement in hardness enhanced the wear resistance of the aluminum alloy. Guo and Caslaru [32] fabricated and characterized micro dent arrays produced on Ti–6Al–4V surfaces. They reported that enhanced laser power intensity resulted in a noticeable effect on surface roughness, which further influenced the frictional characteristics of Ti-6Al-4V surfaces. Furthermore, it was revealed that owing to strain hardening, the peened surface exhibited enhanced micro-hardness. They also analyzed the impact of laser power intensity on surface topography by producing micro indentations at 1, 2, and 3 W. Figure 7.7 represents the microstructure of the indentation produced on Ti-6Al-4V samples. It was found that the dent depth increases with laser power. A depth of 1 μm was obtained for a laser power of 3 W.

The number of laser pulses is another process parameter that influences the size, shape, and depth of the laser-textured surface. Vilhena et al. [33] revealed that the geometry of crater developed on 100Cr6 steel by the LST process was severely influenced by the number of laser pulses. They further reported that the size of depth, height, and diameter of crater can be varied by selecting appropriate processing conditions. Yang et al. [34] reported that with an increase in the number of pulses from 10 to 90, the radius of crater enhanced from 12.72 to 13.37 μm for the bearing steel sample. Semaltianos et al. [35] investigated the effect of laser power and scan speed while modifying surface nickel-based superalloy by laser surface texturing technique. They revealed that pores formation on the sample surface during the LST process can be avoided by using lower power intensity combined with high scan speed and a higher number of over scans.

Lei et al. [26] developed micro-holes on the rake face of tungsten carbide cutting inserts by using femtosecond laser. To develop textures, WC inserts were kept on a translation stage as shown in Figure 7.8 and, with a computer control, their movement is controlled. During their experiment, they varied pulse energy and time duration to

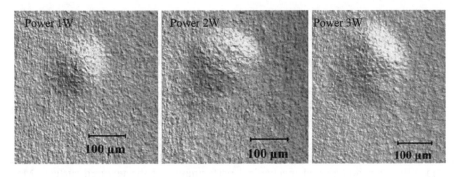

FIGURE 7.7 Surface topography of Ti-6Al-4V processed laser intensities of 1, 2, and 3 W [32].

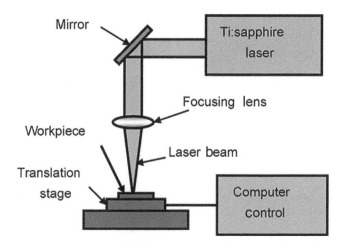

FIGURE 7.8   Experimental setup for femtosecond laser texturing [26].

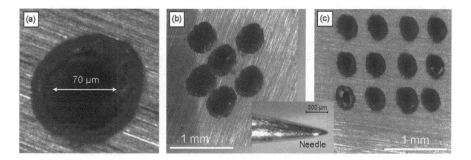

FIGURE 7.9   (a–c) Optical images of micro-holes fabricated WC inserts [26].

control the depth of hole. The main objective was to fabricate deep holes so that they could encompass more cutting fluid/lubricant while cutting. It was found that even at a short pulse duration value the heat-affected zone is apparent owing to the continual laser ablation at a single point for an extended time (~60 seconds). Figure 7.9a–c shows the micro-holes fabricated on WC cutting inserts by femtosecond laser texturing. They performed a finite element analysis (FEM) to study the effect of micro-holes developed on the mechanical performance of WC inserts. It was found that there is no severe stress concentration on the rake face of the tool owing to the holes fabricated by LST. It was further reported that the experimental results supported the FEM analysis due to the excellent strength of the tool while cutting.

Kawasegi et al. [36] fabricated a groove 2.2 m wide and 1.3 m deep by laser ablation with 50 mW power and 250 m/s scan speed on a cemented carbide tool. They developed three different kinds of textures, such as linear texture perpendicular, parallel, and cross texture in the direction of chip flow to analyze the effect of texture on the friction between tool and chip. Schematic representation of textures produced in variation directions on cemented carbide tool is illustrated in Figure 7.10.

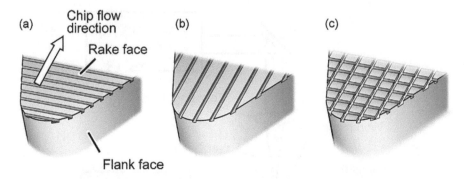

FIGURE 7.10   (a–c) Schematic illustration of texture direction perpendicular, parallel, and cross texture in the direction of chip flow [36].

Sugihara et al. [4] developed a laser surface-textured cubic boron nitride cutting tool with texturing on the flank face for the machining of Inconel 718. They pointed out that the optimum dimensions and shapes of textures depend on the workpiece to be machined, and the appropriate textures to be developed for the machining of Inconel 718 are still unknown. They developed periodic rectangular grooves with 5 μm depth and 20 μm width on a CBN cutting tool using a femtosecond with a wavelength of 800 nm, pulse width of 150 fs, cyclic frequency of 1 kHz, and pulse energy of 300 μJ. Mishra et al. [27] fabricated chevron-shaped textures on the rake face of the tool using an Nd: YAG nanosecond laser. The parameters employed to generate textures are wavelength of 1064 nm, pulse duration of 20 ns, pulse energy of 25 mJ, frequency 20 kHz, scanning speed of 50 mm/s, and pressure of 3 bar. They fabricated textures at a distance of 100 μm from the cutting edge to avoid failure of cutting edge. Sasi et al. [29] fabricated a micro-scale dimple texture on the rake face of the HSS cutting tool using laser texturing to machine Al7075-T6 aerospace alloy. It was found that to fabricate visible dimples, a beam energy of 40 mW/cm^2 and exposure time of 40 seconds is minimum required. They reported that at 100 seconds exposure time a reasonable amount of carbon deposition was observed, thus, during their experiment, exposure time was fixed at a maximum of 90 seconds. Further, in their experiment, beams with wavelengths of 1064, 532, and 355 nm were employed. The scanning electron microscopy (SEM) images of dimples developed at a wavelength of 1064 nm are represented in Figure 7.11.

From the SEM image, the presence of radial flow lines can be witnessed inside the dimple owing to the rapid cooling and solidification of molten metal. Further, cracks can be observed on the surface of dimples owing to the re-deposition of molten metal. In this work, a detailed analysis of the effect of process parameters on the fabrication of dimples on cutting tools by laser texturing was discussed, and it was revealed that wavelength is the most significant factor in laser ablation. Ze et al. [28] investigated the performance of surface textures generated using Nd:YAG laser system on the rake face of self-lubricated cemented carbide (WC/Co) inserts while machining Ti-6Al-4V alloy. To fabricate textures on the rake face, they used an NDYAG laser with a wavelength of 1064 nm, repetition rate of 2 kHz, and a pulse duration of about 20 ns. Further, parameters such as voltage of 12 V, processing current of

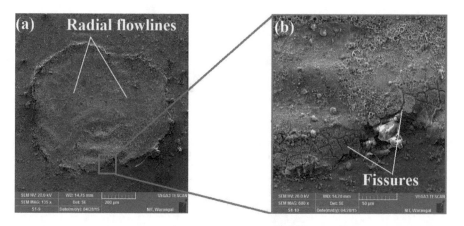

FIGURE 7.11 SEM image of (a) a dimple fabricated at 1064 nm, 100 mW/cm², and 70 seconds (b) magnified image of fissures [29].

20 A, and the processing speed of 10 mm/s were kept constant. It was reported that the performance and effectiveness of a cutting tool mainly depend on the machining conditions. Cutting forces and temperature reduction using textured tools are more obvious at low cutting speed and a relatively higher feed rate.

7.4.2 Texturing Methodology

Surface modification by the LST technique can be achieved mainly using three methods, i.e., laser-induced ablation, patterns generation by laser shock-induced deformation, and melting of material by laser interference. Direct laser ablation is one primary laser surface texturing technique to remove materials from material surface. In this process, when the laser exposes to the target surface, it absorbs energy and causes evaporation of material from the surface. Evaporated material further ionizes and forms plasma. In this way, material can be removed from sample surfaces and grooves or craters of specific geometry can be fabricated. The schematic representation of LST by laser ablation is shown in Figure 7.12.

Laser shock processing (LSP) is an advanced LST technique where shockwaves are used to create compressive residual stress on the material surface. The induced compressive residual stress further induces plastic deformation on the surface of materials. The schematic representation of the fabrication of dent arrays on the material surface is shown in Figure 7.13. In this process, a layer of ablative material is used as a sacrificial layer on the material surface. When the sacrificial layer is exposed to laser light, it vaporizes and forms plasma. The expanding plasma deforms the surface of workpiece material. Direct laser interference patterning (DLIP) or laser interference is another laser surface texturing technique where patterns can be fabricated on materials by the interaction of laser beams. In this process, workpiece material melts in location where maximum interference occurs with laser intensity. Further, materials from these interference maxima relocate to minima locations due to the surface tension phenomenon owing to the temperature differences [30]. Liu et al. [37] investigated the surface modification of Ti-6Al-4V alloy using two-beam laser

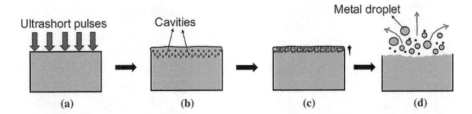

FIGURE 7.12 (a) Laser exposed to material surface, (b) formation of cavities, (c) movement of cavities toward the top, and (d) removal of material by evaporation [38].

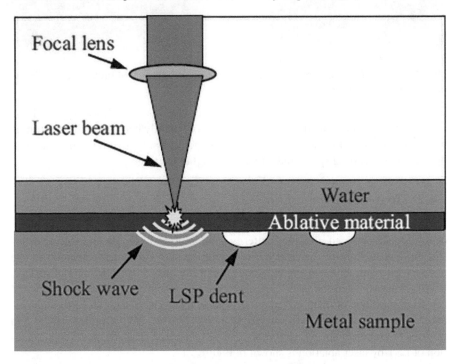

FIGURE 7.13 Schematic representation of fabrication of dent arrays on material surface [32].

interference. In their work, groove structures were fabricated on Ti-6Al-4V surface which enhanced the surface roughness up to 1 µm and abridged the contact angle to approximately 44°.

Kümmel et al. [39] adopted laser texturing on the rake faces of cemented carbide cutting tools by Yb fiber laser system. To fabricate dimples on cutting tools, a laser power of 17.6–19.5 W and a frequency of 20 kHz were used. After texturing, some debris was present on the tool surface owing to the thermal effect, which was later removed by polishing with SiC emery paper combined with final polishing using diamond paste. SEM images of different textures developed on the rake face of the tool are indicated in Figure 7.14. The rectangular box marked in the image is the main cutting edge of the cutting tool. Pacella and Brigginshaw [40] improved the

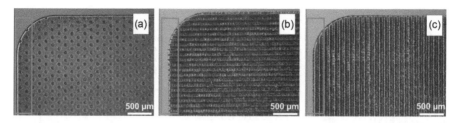

FIGURE 7.14   Different textures developed on the rake face of the tool: (a) dimple texture, (b) channels perpendicular to the cutting edge, (c) channels parallel to the cutting edge [39].

performance of boron nitride cutting tools by using laser micromachining in order to perform dry turning. To process the samples, fiber laser with 1060 nm wavelength with a maximum output power of 70 W was used. Laser machine was CNC controlled and functioned in pulse mode. They reported that peak power was 70 kHz, the pulse duration was 46 ns, and the maximum pulse energy was 1 mJ, respectively, as the optimized condition for laser micromachining.

7.4.3   Texture Optimization

Costes et al. [41] investigated the influence of composition on the cubic boron nitride tool, i.e., grain size and binder material on its performance (tool life, tribology), and highlighted the importance of optimizing such parameters while machining materials like Inconel 718 and Ti-6Al-4V alloy. The above-highlighted problems can be solved by surface texturing the tool material. However, the optimization of texturing process parameters is very important to achieve the finest performance. A general guideline is not yet available for the optimization of the texture design, thus experimental investigation is important to determine the appropriate process parameters to achieve preeminent performance.

Over the years, many researchers have investigated laser surface texturing on cutting tools, and it has been found that optimal texture shapes and dimensions of the textures vary with the materials to be machined and cutting conditions. Sugihara et al. [4] developed a cubic boron nitride (CBN) cutting tool with a textured flank face for the machining of Inconel 718. In their investigation, a detailed analysis was conducted on CBN tool failure during the machining of Inconel 718 to find out the effective texture on the surface of the tool. The schematic representation of the machining of Inconel 718 using the CBN tool is illustrated in Figure 7.15.

To investigate the impact of groove direction on the CBN cutting tool, three different types of texturing were fabricated. The one with microgrooves orthogonal to the main cutting edge (CBN-OR), and the second one with microgrooves orthogonal to the cutting edge but 30 μm from the cutting edge (CBN-ORE) and microgrooves parallel to the main cutting edge (CBN-PA). SEM images of different tools after laser texturing are represented in Figure 7.16. After a number of experiments, it was revealed that microgrooves fabricated orthogonal to the cutting edge of a CBN tool with 30 μm away from the edge (CBN-ORE) considerably reduced the tool wear compared to traditional CBN tool without laser surface texturing. This further reveals

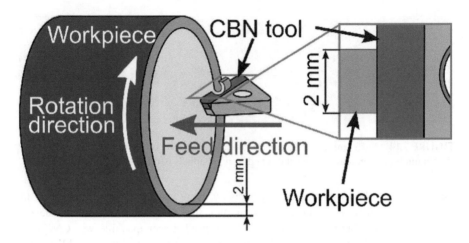

FIGURE 7.15 Schematic representation of machining of Inconel 718 using CBN tool [4].

that the fabrication of microgrooves on flank face of the CBN tool by laser surface texturing is a promising technique for the dry machining of Inconel 718.

Kawasegi et al. [36] fabricated micro and nano-textures on cemented carbide tool owing to its high-speed patterning by femtosecond laser. They examined the impact of irradiation conditions on the final shape and structure of texture developed. Further, they investigated the effects of scanning speed and output power on the shape of textures developed. Figure 7.17 represents the AFM topography image of textured cemented carbide tool. Figure 7.17a shows a groove of 2.2 m wide and 1.3 m deep produced at an output power of 50 mW and a scanning speed of 250 m/s by the traditional laser ablation method. Similarly, Figure 7.17b and c represents the textures produced by laser interferometry method with 100 mW power and 500 m/s scanning speed by scanning perpendicular and parallel to the polarization directions. From the images, it was revealed that texture direction can be controlled by the polarization direction. After a detailed investigation, it was revealed that the shape of the texture had minimal impact at lower cutting speed owing to the adhesion of work material with the tool. However, machining at a higher cutting speed the amount of material adhered to the tool was negligible. Further, from the machining, it was observed that straight chips were produced for using tools with perpendicular waviness, while for parallel textured tool, wavy chips were formed. It was also found that texture shape had an effect on cutting force and the cutting force decreased with increase in depths up to 2.9 µm, but cutting forces augmented when depth is above 2.9 µm. Thus, it was found that a depth of 2.9 µm is apt for texture shape development.

7.4.4 TRIBOLOGY OF TEXTURED SURFACES

Coefficient of friction (CoF) is a significant factor as far as the tribological performance of the textured surface is concerned. In the LST process, by controlling the geometrical aspects of material surfaces, the CoF value can be varied [30]. Many researchers investigated the friction coefficient of laser textured surfaces under dry

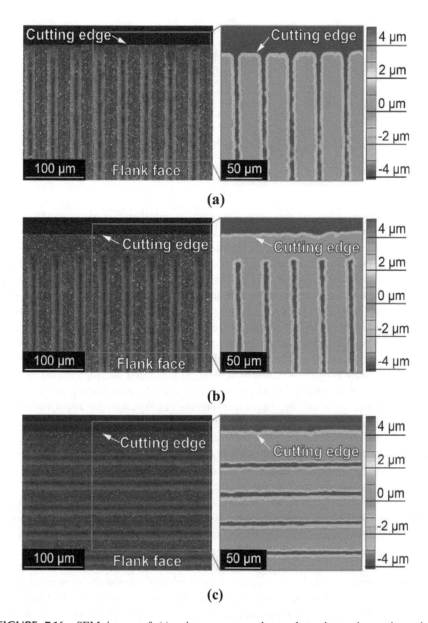

FIGURE 7.16 SEM image of (a) microgrooves orthogonal to the main cutting edge (CBN-OR), (b) microgrooves orthogonal to cutting edge with setback from the cutting edge (CBN-ORE), and (c) microgrooves parallel to the main cutting edge (CBN-PA) [4].

conditions. Wu et al. [42] evaluated the tribological characteristics of laser surface-textured Ti alloys under dry sliding contact. It was revealed that the friction coefficient of textured titanium surfaces filled with molybdenum disulfide (MoS_2) lubricant was reduced up to 40% compared to untextured surfaces. Li et al. [43] performed LST and fabricated dimples with 150 μm on nickel-based composites by Nd: YAG pulsed laser.

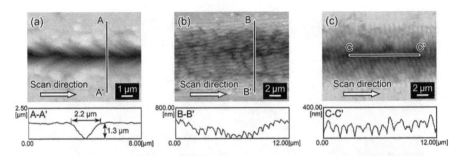

FIGURE 7.17 (a–c) AFM topography image of textured cemented carbide tool [36].

They performed a ball-on-disc wear test with an alumina disc as counterface with 0.4 m/s sliding speed and 20–100 N normal load. It was reported that the friction coefficient of nickel-based composites reduced from 0.18 to 0.1 after texturing.

Several researchers investigated the friction coefficient of laser textured surfaces under lubricated conditions. Wan et al. [44] conducted tribological experiments using a ring-on-disc tester to analyze the frictional performance of laser textured T8 steel discs. It was found that CoF decreased to 0.12 from 0.42 owing to the micropores created by laser which reduced the contact area, facilitated entrapment of wear particles, and reduced plowing. Hsu et al. [45] examined the effect of textural patterns on friction under lubricated conditions. They reported that texture geometry that helps in the prompt changeover to hydrodynamic lubrication, wedged implanted into the dimples, generation of cavitation lift force using reverse flow, etc. can reduce the friction coefficient.

Wear resistance is another important tribological aspect as far as improvements in surface properties are concerned. Wu et al. [42] examined the wear characteristics of textured and untextured Ti alloys filled with MoS_2 solid lubricants, sliding against GCr15 steel balls with the hardness of 60 HRC. It was reported that textured Ti alloys exhibited improved wear resistance owing to the formation of thin lubricating film at the interface of sample and steel balls. In recent times, Yuan et al. [46] evaluated the wear behavior of modified Ti-6l-4V alloy using double-glow plasma surface molybdenizing combined with laser surface texturing technique. Textured Ti-6Al-4V alloy possessed a lower specific rate as compared to untextured one after dry sliding against Si_3N_4 under a normal load of 10 N.

Sugihara et al. [4] investigated the performance of laser-textured CBN tool while machining Inconel 718. It was found that tools with microgrooves orthogonal to the main cutting edge (CBN-OR) and parallel to the cutting edge (CBN-PA) had poor tool life during high-speed machining of Inconel 718. However, for tools with microgrooves orthogonal to cutting edge with setback from the cutting edge (CBN-ORE), even after 200 m cutting its flank face was covered with an adhesion layer, and no detaching of adhesion layer was observed (as represented in Figure 7.18). Therefore, it is very clear that microgrooves alleviated the adhesion layer on the flank face of the CBN tool and further enhanced the tool life. This research work revealed that the CBN-ORE tool had excellent wear resistance and highlighted the importance of proper texturing to enhance the mechanical performance of the CBN tool for the high-speed cutting of Inconel 718.

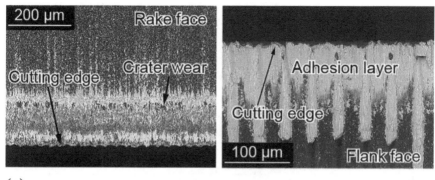

(a)

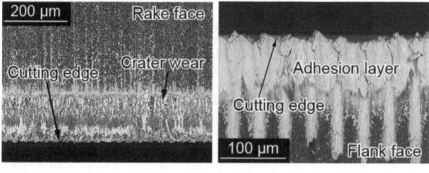

(b)

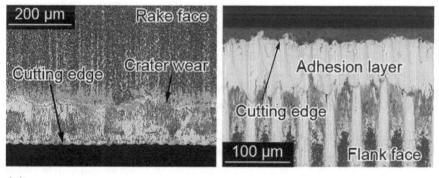

(c)

FIGURE 7.18 SEM image of tools with microgrooves orthogonal to cutting edge with set-back from the cutting edge (CBN-ORE) after (a) 100 m, (b) 150 m, and (c) 200 m cutting [4].

Ze et al. [28] compared the tool life and wear of conventional tool and laser tex-tured cemented carbide (WC/Co) inserts filled with MoS_2 lubricants in the micro-holes while machining Ti-6Al-4V alloy. SEM image of rake face of the conventional tool and laser textured tool after dry cutting of Ti alloy is illustrated in Figure 7.19.

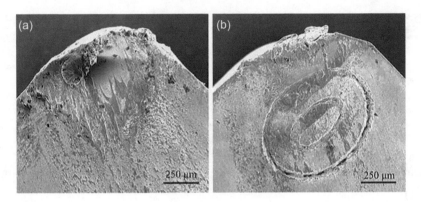

FIGURE 7.19   SEM image of rake face of (a) conventional and (b) laser-textured tool after dry cutting of Ti-6Al-4V [28].

From the image (Figure 7.19a) it is clear that significant wear in form of adhesion and abrasion happened on the conventional tool. However, the textured tool exhibited good wear resistance at the same machining conditions (Figure 7.19b). From the results, it can be inferred that the solid lubricant from the grooves acted as a thin film on the rake face of the textured cutting tool and diminished the tool wear during the cutting process.

Kawasegi et al. [36] discussed the effect of micro-scale and nano-scale textures on cemented carbide tool on its frictional characteristics while high-speed machining. It was reported that the cutting force required for machining with the textured tool is less as compared to nontextured cutting tool. It was revealed that the reduction in cutting force is due to the texture which reduced the friction on the rake face of tool. The reduction in friction is due to the difference in contact length between textured tool and workpiece material.

To improve the friction characteristics and to reduce the adhesion between the rake face of tool and workpiece, laser textures were developed on cemented carbide tool by Xing et al. [47] for the dry cutting of aluminum alloys. During the cutting operation, they investigated the performance of cutting tool in terms of cutting forces, friction coefficient, and tool adhesion. It was revealed that the performance of the laser-textured cutting tool was enhanced at low cutting speeds. Further, it was found that, among the three texture patterns, i.e., linear, circular, and rectangular, tool with rectangular shaped textures is more effective as compared to other tools. It was reported that the coefficient of friction initially increases and then decreases with a rise in cutting velocity; however while machining with the textured tool, the coefficient of friction associated with the tool is reduced, which in turn enhanced the lubricity. On the other hand, for nontextured tools, chips are pressed toward the rake face of the cutting tool and increased the contact area and resulted in high friction coefficient and greater cutting forces. The optical images of rake face of conventional tool (CT), tools with linear texture (TT-L), circular texture (TT-C), and rectangular texture (TT-R) after machining 11.7 and 87.8 m at 54.9 m/min is illustrated in Figure 7.20. It was revealed that textured tools reduced the contact length between tool and chip in comparison with conventional nontextured tools at 11.7 m

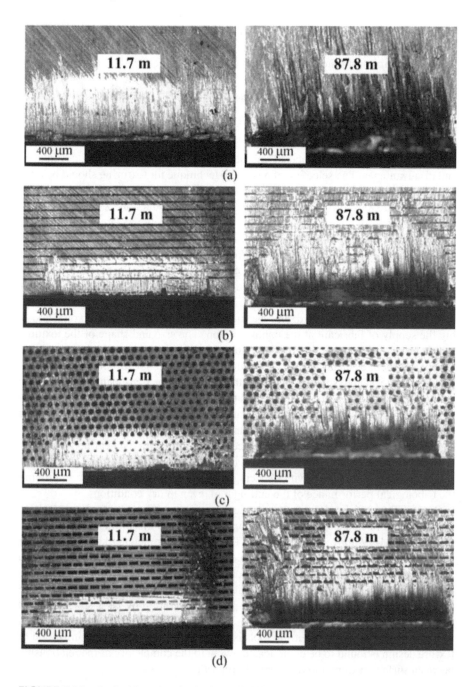

FIGURE 7.20 Optical images of rake face of different tools after dry cutting [47]. (a) CT, (b) TT-L, (c) TT-C, (d) TT-R.

dry cutting. After 87.8 m of cutting, a significant amount of workpiece material has adhered to the rake face of the tool which can be observed from the optical images. At the same time, adhesion on the textured tool is significantly less as compared to the conventional tool. Further, rectangular textured toll exhibited exceptional antiadhesion behavior as compared to other tools.

7.5 CONCLUSIONS

Surface texturing is a multifaceted approach; nevertheless, it has many advantages and disadvantages. The selection of a specific technique for texturing should be chosen based on dimensional accuracy, high surface integrity (fewer defects), enhanced production rate, cost efficiency, and optimum rate of production. A superior reduction in tool wear (flank and crater), cutting forces, and frictional forces is achievable through texturing of cutting tools. Providing textures on the cutting tool surfaces can be an advantage for dry and wet cutting/machining applications. In both cases, the contact area textures assist in reducing the contact area between the tool and the workpiece surfaces. During dry machining, the wear debris will be trapped in the vacant spaces within the textures and can lead to built-up stabilization. For wet machining, the lubricant will be entrapped, and the surface dimples can act as reservoirs, assisting the supply of lubricant at the contact region. The size and shape of the textures should be optimized such that they can hold a sufficient quantity of lubricants to be supplied at the interfacial zone. The geometrical/dimensional parameters of the textures, viz. size, shape, spacing, and orientation, which are significantly governed by the cutting process/conditions and the material (tool/workpiece) should be properly optimized for maximum performance. Parallel placement of textures at the cutting edge can aid in reducing friction and wear caused by the chip formation and associated mechanisms. Nevertheless, their placement should be in the near neighborhood of the cutting edge rather than exactly at the cutting edge. Besides, the inclusion of solid lubricants within the cavities/spacing within the textures can aid in improving the tribological performance of the cutting tool, even in dry conditions.

The LST technique can be used to develop microgrooves on tool materials to improve their mechanical performance during high-speed machining of hard materials like Ti-6Al-4V alloy and Inconel 718. The process parameters like spot size, laser power, scan speed, energy input, and feed rate can control the quality of textured surfaces. The effect of different parameters on laser surface texturing is discussed. Surface modification by the LST technique can be achieved mainly using three methods, such as laser-induced ablation, pattern generation by laser shock-induced deformation, and melting of material by laser interference. Further, the enhanced wear resistance of textured cutting tools is also discussed here. The optimization of texture design is found to be complicated owing to the complexity of the relationship between surface texture and tribological properties.

7.6 FUTURE SCOPE

At present, the optimization of texture design is complicated owing to the complexity of the relationship between the texture of material and tribological properties, along with mechanical performance, morphology, and microstructure evolution. A numerical or computational modeling can be developed as an alternative to the existing expensive experimental investigation adopted for laser surface texturing optimization.

REFERENCES

1. Ezugwu EO. Key improvements in the machining of difficult-to-cut aerospace superalloys. *Int J Mach Tools Manuf* 2005;45:1353–67. doi:10.1016/j.ijmachtools.2005.02.003.
2. Mohsan AUH, Liu Z, Padhy GK. A review on the progress towards improvement in surface integrity of Inconel 718 under high pressure and flood cooling conditions. *Int J Adv Manuf Technol* 2017;91:107–25. doi:10.1007/s00170-016-9737-3.
3. Krämer A, Lung D, Klocke F. High performance cutting of aircraft and turbine components, 2012:425–32. doi:10.1063/1.4707592.
4. Sugihara T, Nishimoto Y, Enomoto T. Development of a novel cubic boron nitride cutting tool with a textured flank face for high-speed machining of Inconel 718. *Precis Eng* 2017;48:75–82. doi:10.1016/j.precisioneng.2016.11.007.
5. Childs THC, Maekawa K, Obikawa T, Yamane Y. *Metal Machining: Theory and Applications*. Oxford: Butterworth-Heinemann,; 2000.
6. Kumar Mishra S, Ghosh S, Aravindan S. Machining performance evaluation of Ti6Al4V alloy with laser textured tools under MQL and nano-MQL environments. *J Manuf Process* 2020;53:174–89. doi:10.1016/j.jmapro.2020.02.014.
7. Song W, Wang Z, Deng J, Zhou K, Wang S, Guo Z. Cutting temperature analysis and experiment of Ti–MoS₂/Zr-coated cemented carbide tool. *Int J Adv Manuf Technol* 2017;93:799–809. doi:10.1007/s00170-017-0509-5.
8. Wang Y, Li C, Zhang Y, Yang M, Li B, Dong L, et al. Processing characteristics of vegetable oil-based nanofluid MQL for grinding different workpiece materials. *Int J Precis Eng Manuf Technol* 2018;5:327–39. doi:10.1007/s40684-018-0035-4.
9. Wang J, Armarego EJA. Computer-aided optimization of multiple constraint single pass face milling operations. *Mach Sci Technol* 2001;5:77–99. doi:10.1081/MST-100103179.
10. Bouzakis K-D, Michailidis N, Skordaris G, Bouzakis E, Biermann D, M'Saoubi R. Cutting with coated tools: Coating technologies, characterization methods and performance optimization. *CIRP Ann* 2012;61:703–23. doi:10.1016/j.cirp.2012.05.006.
11. Wakuda M, Yamauchi Y, Kanzaki S, Yasuda Y. Effect of surface texturing on friction reduction between ceramic and steel materials under lubricated sliding contact. *Wear* 2003;254:356–63. doi:10.1016/S0043-1648(03)00004-8.
12. Ryk G, Kligerman Y, Etsion I. Experimental investigation of laser surface texturing for reciprocating automotive components. *Tribol Trans* 2002;45:444–9. doi:10.1080/10402000208982572.
13. Vencl A, Ivanovic L, Stojanovic B, Svoboda P. Surface texturing for tribological applications: A review. *SERBIATRIB '19 16th International Conference on Tribology*, Kragujevac, Serbia; 2019, pp. 227–39.
14. Yu H, Deng H, Huang W, Wang X. The effect of dimple shapes on friction of parallel surfaces. *Proc Inst Mech Eng Part J J Eng Tribol* 2011;225:693–703. doi:10.1177/1350650111406045.
15. Rahmani R. An Investigation into analysis and optimisation of textured slider bearings with application in piston ring/cylinder liner contact. Anglia Ruskin University, 2008.
16. Schneider J, Braun D, Greiner C. Laser Textured Surfaces for Mixed Lubrication: Influence of Aspect Ratio, Textured Area and Dimple Arrangement. *Lubricants* 2017;5:32. doi:10.3390/lubricants5030032.

17. Rathod P, Aravindan S, Paruchuri VR. Evaluating the effectiveness of the novel surface textured tools in enhancing the machinability of titanium alloy (Ti6Al4V). *J Adv Mech Des Syst Manuf* 2015;9:JAMDSM0035–JAMDSM0035. doi:10.1299/jamdsm.2015jamdsm0035.

18. Ma J, Ge X, Qiu C, Lei S. FEM assessment of performance of microhole textured cutting tool in dry machining of Ti-6Al-4V. *Int J Adv Manuf Technol* 2016;84:2609–21. doi:10.1007/s00170-015-7918-0.

19. Arulkirubakaran D, Senthilkumar V, Kumawat V. Effect of micro-textured tools on machining of Ti–6Al–4V alloy: An experimental and numerical approach. *Int J Refract Met Hard Mater* 2016;54:165–77. doi:10.1016/j.ijrmhm.2015.07.027.

20. Nouri A, Wen C. Introduction to surface coating and modification for metallic biomaterials. *Surf Coat Modif Met Biomater*, Elsevier; 2015, pp. 3–60. doi:10.1016/B978-1-78242-303-4.00001-6.

21. Singh A, Harimkar SP. Laser surface engineering of magnesium alloys: A review. *JOM* 2012;64:716–33. doi:10.1007/s11837-012-0340-2.

22. Kennedy E, Byrne G, Collins DN. A review of the use of high power diode lasers in surface hardening. *J Mater Process Technol* 2004;155–156:1855–60. doi:10.1016/j.jmatprotec.2004.04.276.

23. Etsion I. Improving tribological performance of mechanical seals by laser surface texturing. *Proceedings of the International Pump Users Symposium*, Texas; 2000, p. 17–22.

24. Brizmer V, Kligerman Y, Etsion I. A laser surface textured parallel thrust bearing. *Tribol Trans* 2003;46:397–403. doi:10.1080/10402000308982643.

25. Howes TD, Tönshoff HK, Heuer W, Howes T. Environmental aspects of grinding fluids. *CIRP Ann* 1991;40:623–30. doi:10.1016/S0007-8506(07)61138-X.

26. Lei S, Devarajan S, Chang Z. A study of micropool lubricated cutting tool in machining of mild steel. *J Mater Process Technol* 2009;209:1612–20. doi:10.1016/j.jmatprotec.2008.04.024.

27. Mishra SK, Ghosh S, Aravindan S. Characterization and machining performance of laser-textured chevron shaped tools coated with AlTiN and AlCrN coatings. *Surf Coatings Technol* 2018;334:344–56. doi:10.1016/j.surfcoat.2017.11.061.

28. Ze W, Jianxin D, Yang C, Youqiang X, Jun Z. Performance of the self-lubricating textured tools in dry cutting of Ti-6Al-4V. *Int J Adv Manuf Technol* 2012;62:943–51. doi:10.1007/s00170-011-3853-x.

29. Sasi R, Kanmani Subbu S, Palani IA. Performance of laser surface textured high speed steel cutting tool in machining of Al7075-T6 aerospace alloy. *Surf Coatings Technol* 2017;313:337–46. doi:10.1016/j.surfcoat.2017.01.118.

30. Mao B, Liao Y, Li B. Gradient twinning microstructure generated by laser shock peening in an AZ31B magnesium alloy. *Appl Surf Sci* 2018;457:342–51. doi:10.1016/j.apsusc.2018.06.176.

31. Sánchez-Santana U, Rubio-González C, Gomez-Rosas G, Ocaña JL, Molpeceres C, Porro J, et al. Wear and friction of 6061-T6 aluminum alloy treated by laser shock processing. *Wear* 2006;260:847–54. doi:10.1016/j.wear.2005.04.014.

32. Guo YB, Caslaru R. Fabrication and characterization of micro dent arrays produced by laser shock peening on titanium Ti–6Al–4V surfaces. *J Mater Process Technol* 2011;211:729–36. doi:10.1016/j.jmatprotec.2010.12.007.

33. Vilhena LM, Sedláček M, Podgornik B, Vižintin J, Babnik A, Možina J. Surface texturing by pulsed Nd:YAG laser. *Tribol Int* 2009;42:1496–504. doi:10.1016/j.triboint.2009.06.003.

34. Yang L, Ding Y, Cheng B, He J, Wang G, Wang Y. Investigations on femtosecond laser modified micro-textured surface with anti-friction property on bearing steel GCr15. *Appl Surf Sci* 2018;434:831–42. doi:10.1016/j.apsusc.2017.10.234.

35. Semaltianos NG, Perrie W, French P, Sharp M, Dearden G, Watkins KG. Femtosecond laser surface texturing of a nickel-based superalloy. *Appl Surf Sci* 2008;255:2796–802. https://doi.org/10.1016/j.apsusc.2008.08.043.

36. Kawasegi N, Sugimori H, Morimoto H, Morita N, Hori I. Development of cutting tools with microscale and nanoscale textures to improve frictional behavior. *Precis Eng* 2009;33:248–54. doi:10.1016/j.precisioneng.2008.07.005.

37. Liu Q, Li W, Cao L, Wang J, Qu Y, Wang X, et al. Ti-6Al-4V alloy modification by laser interference lithography. *2016 IEEE International Conference on Manipulation, Manufacturing and Measurement.* Nanoscale, IEEE, Chongqing, China; 2016, pp. 366–70. doi:10.1109/3M-NANO.2016.7824944.

38. Maharjan N, Zhou W, Zhou Y, Guan Y. Ablation morphology and ablation threshold of Ti-6Al-4V alloy during femtosecond laser processing. *Appl Phys A* 2018;124:519. doi:10.1007/s00339-018-1928-3.

39. Kümmel J, Braun D, Gibmeier J, Schneider J, Greiner C, Schulze V, et al. Study on micro texturing of uncoated cemented carbide cutting tools for wear improvement and built-up edge stabilisation. *J Mater Process Technol* 2015;215:62–70. doi:10.1016/j.jmatprotec.2014.07.032.

40. Pacella M, Brigginshaw D. Enhanced wear performance of laser machined tools in dry turning of hardened steels. *J Manuf Process* 2020;56:189–96. doi:10.1016/j.jmapro.2020.04.058.

41. Costes JP, Guillet Y, Poulachon G, Dessoly M. Tool-life and wear mechanisms of CBN tools in machining of Inconel 718. *Int J Mach Tools Manuf* 2007;47:1081–7. doi:10.1016/j.ijmachtools.2006.09.031.

42. Wu Z, Xing Y, Huang P, Liu L. Tribological properties of dimple-textured titanium alloys under dry sliding contact. *Surf Coatings Technol* 2017;309:21–8. doi:10.1016/j.surfcoat.2016.11.045.

43. Li J, Xiong D, Dai J, Huang Z, Tyagi R. Effect of surface laser texture on friction properties of nickel-based composite. *Tribol Int* 2010;43:1193–9. doi:10.1016/j.triboint.2009.12.044.

44. Wan Y, Xiong D-S. The effect of laser surface texturing on frictional performance of face seal. *J Mater Process Technol* 2008;197:96–100. doi:10.1016/j.jmatprotec.2007.06.019.

45. Hsu SM, Jing Y, Hua D, Zhang H. Friction reduction using discrete surface textures: Principle and design. *J Phys D Appl Phys* 2014;47:335307. doi:10.1088/0022-3727/47/33/335307.

46. Yuan S, Lin N, Zou J, Liu Z, Yu Y, Ma Y, et al. Manipulation tribological behavior of Ti6Al4V alloy via a duplex treatment of double glow plasma surface molybdenizing-laser surface texturing (LST). *J Mater Res Technol* 2020;9:6360–75. doi:10.1016/j.jmrt.2020.03.061.

47. Xing Y, Deng J, Wang X, Ehmann K, Cao J. Experimental assessment of laser textured cutting tools in dry cutting of aluminum alloys. *J Manuf Sci Eng* 2016;138. doi:10.1115/1.4032263.

8 Surface Engineering
Coatings and Surface Diagnostics

Jyoti Menghani, Rajat Thombare,
and Saravanan Ramesh
SVNIT

CONTENTS

This chapter is mainly divided into two classes: surface engineering techniques and surface characterization. In surface engineering technique, the emphasis is laid on metallic coating and metal substrate, while the latter class covers the widely used sophisticated techniques for surface characterization.

DOI: 10.1201/9781003097082-8

SURFACE ENGINEERING TECHNIQUES

8.1 INTRODUCTION

The field of material sciences, since the very moment of its birth, has focused on the bulk matter found around us and the modification of the same. Not much attention was paid to the surface properties since the field was saturated with new and exciting materials such as plastics, composites, and light metals like aluminum and its alloys, and other rare earth metals that proved to be excellent alloying metals. However, a saturation gradually arose in the discovery of new materials and, hence, forced scientists and engineers to focus not only on the bulk properties of the materials but also on their surface properties.

Also, the fact that the engineering failures caused by corrosion, wear, and fatigue occur primarily at the surface and that these account for the loss of plant efficiency, due to shutdowns, and the evidential direct and indirect losses incurred (which still account for billions of dollars annually worldwide), necessitated the need to improve the surface properties. Additionally, for various reasons, including the need for unique properties (biocompatibility, self-lubricating coating, electrical, magnetic, adhesive, ablation, passivation, inhibition, catalytic, and diffusion), lightweight, design and engineering flexibility, materials conversion, or economics, surface modifications or protective surface coatings (thermal barrier coatings, environmental barrier coatings, and mechanical barrier coatings) became desirable and, at times, even necessary. To achieve this purpose, bulk material properties had to be differentiated from surface properties, and thus, a new concept of surface engineering emerged [1–3].

8.1.1 SURFACE ENGINEERING

One of the most effective means of mitigating damage and achieving desired properties is to treat or "engineer" the surface so that it can perform functions that are not served by the bulk of the material [1]. The definition of surface engineering includes the design of substrate and surface collectively as a system to give cost-effective performance enhancement [2].

8.1.2 IMPORTANCE OF SURFACE ENGINEERING

- Versatility and Cost Reduction:
  Surface engineering attempts to produce tools, machine parts, and whole assemblies from low-cost materials of lower properties by imparting their surfaces with enhanced characteristics. Almost all types of materials, including metals, ceramics, polymers, and composites, can be coated with similar or dissimilar materials. It is also possible to form coatings of newer materials (e.g., met glass, b-C_3N_4), graded deposits, multicomponent deposits, etc. [4–6].

  As per reference [7], it is mentioned that the repair of the blades of moderately worn turbines by the application of physical processes of plasma substances does not exceed 25% of the cost of new ones. In addition, a reduction of cost by one-third due to the use of AISI 4140 to replace ISO 41CrAlMo74 as the screw base material and perform laser surface alloying [8]

TABLE 8.1
Industrial Exploitation of Ion Implantation [10]

Material	Application (Specific Examples)	Typical Results
Cemented WC	Drilling (printed circuit board, dental burrs, etc.)	Four times normal life, less frequent breakage, and better end product
Ti6Al4V	Orthopedic implants (artificial hip and knee joints)	Significant (400 times) lifetime increase in laboratory tests
M50, 52100 steel	Bearings (precision bearings for aircraft)	Improved protection against corrosion, sliding wear, and rolling contact fatigue
Various alloys	Extrusion (spanners, nozzles, and dies)	Four to six times normal performance
D2 Steel	Punching and stamping (pellet punches for nuclear fuel, scoring dies for cans	Three to five times normal life

- Improved Reliability

  The reliability and hence service life of machine components can be improved. This would result in reduced breakdowns and failures during operation in a plant and hence increases the production capacity of the machines.

  As per reference [9] depending on the extent of peening imparted, the service life of spring increased by 400%–1200%; in addition, it is mentioned that fatigue life of gears was increased by 500%.

- Energy Losses

  Energy losses due to friction between moving components are reduced due to improved tribological properties achieved by surface engineering.

- Part Replacement

  The frequency of tool replacements and machine overhaul in maintenance is greatly reduced. Surface engineering processes have reduced corrosion losses by 15%–35%, which is of great significance because corrosion losses are as high as 5% of the GDP [1–6].

8.2 CLASSIFICATION OF SURFACE ENGINEERING PROCESSES

The broad classification of surface engineering processes, primarily differentiated based on whether the process modifies the microstructure or composition of the base metal, is shown in Figure 8.1. The methods mentioned in Figure 8.1 are explained in detail in the following section.

8.2.1 MICROSTRUCTURAL MODIFICATION

8.2.1.1 Surface Transformation Hardening

8.2.1.1.1 Transformation Hardening

In the case of ferrous metals like high carbon steels and cast iron, the transformation hardening process consists of heating of the metal surface to the austenitic phase

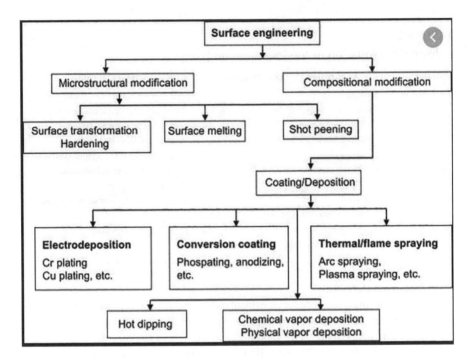

FIGURE 8.1   Classification of surface engineering processes [6].

(the FCC structure) followed by quenching to ambient temperature or subambient temperature to impose metastable transformation of ductile austenite (FCC) to hard martensite (BCT).

a. Laser Transformation Hardening:

In laser transformation hardening, as the name suggests, there is a transformation from austenite to martensite using laser as a heating source. Thin and fine-grained martensitic layer is formed on the surface of substrate due to rapid heating of surface till austenite phase followed by self-quenching (conduction through cooler bulk). Figure 8.2a depicts the principle of laser transformation hardening. The main feature of the process is the precise control of process parameters, very high rates of heating and cooling (10^4 K/s) or greater, and a rapid thermal cycle of the order of 0.1 second. Compared to other surface hardening processes, this process has the advantage of surface hardening of complex and intricate components even after final machining, precise control of hardening zone, minimum distortion, and excellent dimensional tolerance [11–13].

As per reference [14], wherein laser hardening was adopted to improve mechanical and corrosion resistance of AISI P20 plastic die steel, the major advantage of laser surface hardening is that it is not location-sensitive and can be applied in difficult to reach positions unlike the coating process wherein the accurate control of coating thickness and excellent adhesion is

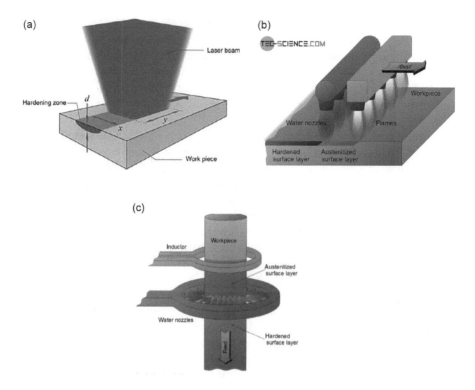

FIGURE 8.2 Surface transformation hardening [18,19]. (a) Laser surface hardening, (b) flame hardening, and (c) induction hardening.

a prime requirement. The results indicated that optimum laser parameters had improved mechanical and corrosion resistance in 3.5% NaCl solution.

b. Flame Hardening:

The surface modification by flame hardening involves rapid heating of the surface up to elevated temperatures by the direct impingement of a high-temperature flame (selection of fuels from acetylene, propane, or natural gas) followed immediately by water quenching. The applications include blades for steam and gas turbines, gears for paper machinery, rolls for the printing and metal-working industries, cams for packaging machinery, machine tool beds, and various automotive components. The major advantages of this process include its simplicity in operation, high speed, local hardening capability, low-cost processing, and the improved mechanical properties achieved. However, flame-hardened steels are more prone to stress corrosion cracking (SCC), brittle fracture, and fatigue failure because the process induces high tensile stress, which is responsible for the degradation of the properties [15]. Figure 8.2b shows the details of flame hardening.

As per reference [16], flame hardening was carried out for low carbon 12Cr steel, which is widely used for low-pressure steam turbine blades in nuclear power plants. The hardness, hardening depth, and residual stresses were determined by varying thermal cycles of flame hardening. The optimum process parameters with minimum residual stress were determined using the temperature cycle, which consists of parameters such as precise control of surface temperature cycles, the exposed height from the water surface, and the cooling rate. It was found that increasing hardening temperature beyond 1200°C increases thermal and transformation stress components. This result was in line with the guidelines given by Siemens.

c. Induction Hardening:

Eddy currents are induced in a metallic workpiece when subjected to a high-frequency alternating current, and this leads to temperature rise due to Joule heating. The heating is immediately followed by either water, oil, or cold air quenching so that the heat does not spread into core of metal, as shown in Figure 8.2c. The process is well suited due to excellent electromagnetic properties of ferrous metals. As with hardening, the surface will exhibit excellent hardness and wear resistance along with tough core and dimensional stability. Further, the process has the advantage of mechanization and minimum contamination [17].

8.2.1.2 Surface Melting

In contrast to surface hardening wherein the surface is heated to the transformation temperature, this process involves local melting of material and allowing it to resolidify. There is no variation in chemical composition but liquid–solid phase transformation occurs. This uses high-power density instruments, as input, like laser or electron-beam heating; electric arc welding method that uses non-consumable electrode and acts as a source of heat, e.g., tungsten inert gas process can also be used. Various modifications in microstructure such as homogenization and refinement, formation of non-equilibrium phases, or glass formation and supersaturation occur due to rapid solidification. Figure 8.3 shows the phenomena occurring during laser surface melting.

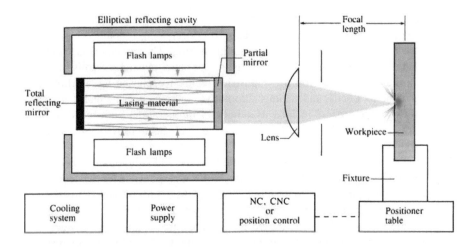

FIGURE 8.3 Laser surface melting [21].

In comparison to other hardening processes, the input of higher power densities is required. The major advantage over the hardening process is that it can be used for both ferrous and nonferrous metals, potentially even for non-metals. As per reference [20], laser surface melting produced finer and cellular microstructures on Mg-Zn alloy and higher treatment depths had better corrosion resistance. In the case of low carbon steels, there is a retention of soft δ-ferrite (due to peritectic reaction) in the quenched hardened surface layer and hence low carbon steels cannot be hardened by laser surface melting [21].

8.2.1.3 Shot Peening

Shot peening is a highly efficient, low cost, simple, and industrially reliable surface treatment process consisting of striking the surface with high-speed pellets (shot made of iron/steel/ceramic-zirconium oxide) and glass beads of roughly 0.2–5 mm diameters and is placed in a chamber at ambient temperatures. Compressive residual stresses are induced which forbid crack initiation and propagation on a component's surface, thus improving fatigue strength.

Shot peening has manifold effects including the formation of a work-hardened layer increasing hardness due to an increase in dislocation density, grain refinement, and phase transformation of metastable austenite to martensite. Figure 8.4 shows the stress distribution of shot peening on the metal surface. It also increases the yield strength but lowers the ductility. Depending on process variables such as velocity, size of shot, shot material, etc., the compressive layer thickness may vary from 0.25 to 0.45 mm.

A variation in shot peening is ultrasonic shock peening, as depicted in Figure 8.4b, wherein the kinetic energy is given by using a vibrational frequency of 20 kHz. Shotless peening involves the use of a plunger pump and water jet peening nozzle. A high-pressurized water jet (20–30 MPa) is directed onto the metal surface, which creates cavitation bubbles and forms a cavitation jet, which then causes surface impact similar to shots. The process has advantages that it is applicable to various ferrous (stainless steels, carburized steels, Hadfield steel, low carbon steels, and high strength steels) and nonferrous (titanium alloys and aluminum alloys) metals [22–24]. As per reference [25], shot peening was performed on two DP600 (ferrite and bainite phase) and FB600 (ferrite, austenite, and martensite) steel grade having high ultimate tensile

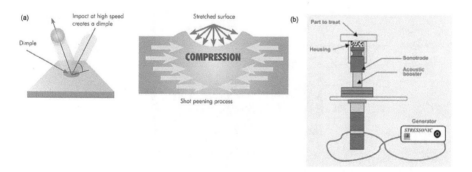

FIGURE 8.4 (a) Shot peening effect [23] and (b) ultrasonic shot peening technique [26].

strength but low hardness, which were hence unsuitable for suspension and chassis components for the automotive industry. Compressive stresses due to shot peening on steel resulted in microstructural changes in terms of grain refinement, which delayed initiation of surface fatigue crack.

8.3 COMPOSITIONAL MODIFICATION (COATING DEPOSITION)

8.3.1 ELECTROCHEMICAL METHODS

Electrochemical methods are the most common and well-established processes for applying the desired coating of metal, oxide, or salt onto the surface of a conductor material by electrolysis process. Here, the potential is applied between the anode (coating material) and the cathode (part to be coated), and a coating is obtained by the reduction of the cations of the anode metal. This process is represented in Figure 8.5. Reduction of ions in solution resulted in the formation of charged particles – positive ion and negative electrons that can cross-interface. Four types of fundamental processes are involved: (1) electrode-solution interface as the locus of the deposition process, (2) kinetics and mechanism of the deposition process, (3) nucleation and growth processes of the deposits, and (4) structure and properties of the deposits.

In addition to decorative applications, this process is widely used as micro and nanofabrication method for a variety of advanced applications. Further, coating deposition is of single or multilayer with thickness varying from 0.3 to 300 μm. The most frequently used coating metals (in ascending order) are cobalt, iron, platinum, rhodium, gold, silver, lead, copper, tin, zinc, nickel, and chromium; and alloys such as Cu-Zn, Co-W, Co-P, Ni-P, Ni-Co, Ni-Fe, Zn-Ni, Sn-Cd, Sn-Ni, and Sn-Pb, used as single or in combination with other coatings, mainly for corrosion protection and for decorative purposes.

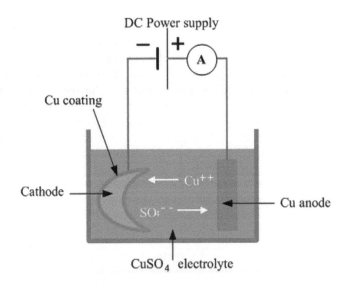

FIGURE 8.5 Electrochemical deposition [30].

Since electrodeposition can be carried out at room temperature, it can be used for the deposition of nanocrystalline coatings. The major advantages include low cost, simple design, high production rate, and the possibility of conversion of laboratory scale to industrial applications. In addition, alloy coating can be carried with a range of elements, and hence more complex products can also be deposited by adjusting direct current parameters or using pulsed current [27,28].

As per reference [29], investigation was carried out on Co deposition of nanocrystalline Ni-Cr coating using direct current and pulse current. The results indicated that the proportion of Cr in NiCr alloy coating is dependent on various pulse current parameters like peak current density, duty cycle (Cr content is directly proportional), and frequency (Cr content is inversely proportional). Due to the change in Cr content, there is a variation in microstructure, and as a result, a change in microhardness and corrosion behavior is observed. Improvement in corrosion resistance was observed up to 24% Cr beyond which there was reduction in corrosion resistance due to the segregation of γ' phase in the Ni–Cr alloy and also due to microcrack formation.

8.3.2 Conversion Coating

As the name suggests, in conversion coatings, a layer of substrate metal oxides, vanadates, chromates, cerate, molybdate, phosphates, or other compounds that are chemically bonded to the surface are produced by a chemical or electrochemical reaction at a metal surface.

Hexavalent chromium conversion coatings are largely used to prevent corrosion of aluminum, zinc, steel, magnesium, cadmium, copper, tin, nickel, silver, Zr, and other substrates, and they improve adhesion of paint or metallic surface for other decorative finishes. The details of the process are shown in Figure 8.6a. The process capability includes its simplicity and high stability. However, due to high toxicity and health risk as per the European Regulation Reach (2013), a ban was adopted on the usage of Cr (VI). Later research was oriented on the development of trivalent chromium process (TCP) in place of Cr (VI)-free coatings. Now, TCP is already commercially available [31–32].

Phosphate coatings or phosphating involves the treatment of the ferrous surface with dilute phosphoric acid and other chemicals where the surface is converted to crystalline phosphate and is widely used because of their low cost of manufacturing, excellent adhesion, high corrosion resistance, better abrasive resistance of the structure, and easy operation. The application of modified phosphate such as zinc phosphate (used for steel fiber reinforced concrete), iron phosphate, and manganese phosphate coatings is for improving the corrosion resistance of steel. Phosphating conversion coatings enhance biocompatibility and osteoconductive of different metallic devices [33–34].

Anodizing of aluminum resulting in the formation of transparent alumina is usually produced using aluminum as an anode in a sulfuric acid bath and by using direct current (DC), as shown in Figure 8.6b. The application of anodizing is in various fields for improving corrosion resistance, enhancing wear resistance, and appealing decorative/cosmetic appearances. Owing to high solar reflectance, white anodized Al surfaces are used in the aerospace industry. The white color of anodized layer can

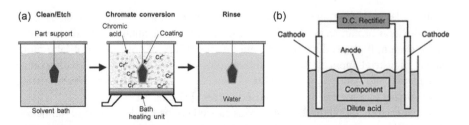

FIGURE 8.6　(a) Chrome conversion coating [38] and (b) anodizing [39].

be varied to different kinds of light gray colors using metal oxide particles in the Al matrix [35].

The latest development in anodizing includes high frequency (HF) pulse anodizing. As per reference [36], anodizing was performed on FSP Al alloy, which was reinforced with TiO_2 particles; the purpose being TiO_2 has high reflectance than Al_2O_3. The FSP surface composite layer was then subjected to HF anodizing in 20 wt% sulfuric acid bath at 10°C, and various anodizing parameters like pulse frequency, pulse duty cycle, and anodic cycle voltage amplitudes were varied to get an anodized layer of desired characteristics. Anodic cycle potential and anodizing frequency determine anodizing film growth, hardness, and total reflectance (growth rate is directly proportional) [37].

8.3.3　Hot Dipping

As the name suggests, the hot dipping process involves the immersion of metal, which is to be coated, into a bath of molten metal. For adequate adherence of coating onto the metal surface (substrate), a brittle intermetallic compound layer of the coating composition and base metal is formed at the surface, the details of which are shown in Figure 8.7. The major applications of hot dipping include alumina and chromia that improve high-temperature oxidation and carburization. Another most common example is galvanizing of steel. The process has the advantages of low cost, simple operation, and large-scale production [40–41].

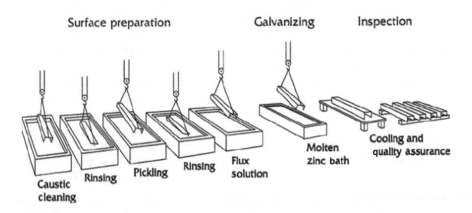

FIGURE 8.7　Hot dip galvanizing [41].

8.3.4 VAPOR DEPOSITION

The broad classification of vapor deposition includes chemical vapor deposition (CVD) and physical vapor deposition (PVD) [42].

a. **Physical Vapor Deposition**

PVD is a coating process and a generalized term to describe a variety of methods to deposit thin films by the condensation of a vaporized form of solid material onto various substrates as thin-film condensed phase [43].

i. **Sputtering:**

In this process, the cathode acts as the target (source of film-forming material) for ion impact sputtering, as shown in Figure 8.8a. The material condenses on substrate through a low-pressure gaseous environment. If the gaseous atmosphere contains reactive gases like nitrogen, carbon, or boron, then respective nitrides, carbides, or borides may form. Such processes have the capability to deposit thin films of elemental, alloy, and compound materials as well as polymeric materials. The process gives a better morphological quality of surfaces where roughness, grain size, stoichiometry, and other requirements are more significant than the deposition rate [44].

ii. **Resistive Heating PVD:**

In this process, the material evaporates or sublimates under vacuum and is deposited as a thin film on the substrate, as shown in Figure 8.8b. The major advantage of this process is its high deposition rate and is most suitable for metals having a low melting point [45].

iii. **Electron Beam PVD:**

Owing to their outstanding properties, such as their strain-tolerant columnar microstructure, good thermal shock resistance, strong bonding to the substrate, and smooth surface of the coatings, the process is widely used for thermal barrier coatings on advanced turbine blades to increase engine efficiency and blade performance. The process involves electron bombardment of the coating material source in a high vacuum, as depicted in Figure 8.8c [46].

iv. **Cathode Arc Deposition:**

In the cathodic arc technique, the parameters for deposition include low voltage and high current between the cathode target (source of metal) and the anode in vacuum to deposit on the substrate. The current density is of the order of 10^6 and $10^{12} A/m^2$, and spot remains on the site for μs or sub-μs time. The process is depicted in Figure 8.8d. The intense energy density leads to transition, at arc spot, from solid metal target to vapor plasma along with a large number of droplets.

The two major drawbacks with cathode arc evaporation are as follows:

1. Presence of macroparticles on the surface, resulting in rough surface and reducing overall performance

2. Restrictions on types of target metal, such as good electroconductivity, high or low melting point, and high mechanical strength.

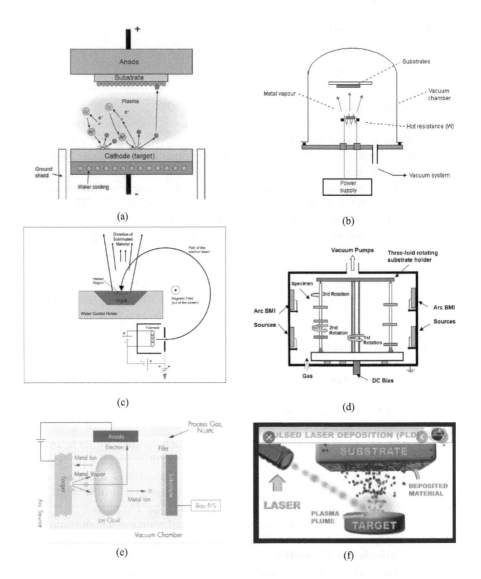

FIGURE 8.8 (a) Magnetron sputtering [50], (b) resistive heating PVD [45], (c) electron beam evaporation [51], (d) cathode arc evaporation [46], (e) ion plating [52], and (f) pulsed laser deposition [53].

However, because of its high target ionization ratio and high deposition rate, cathode arc deposition dominates coating deposition on cutting tools [47]

v. **Ion Plating:**

Ionization of evaporated atoms (plasma) occurs and positive ions are attracted to the negative substrate. On the use of reactive gases, metal carbides and nitrides can be deposited. Details are shown in Figure 8.8e [48].

vi. **Pulsed Laser Deposition:**
Target is struck by a high-power pulsed laser beam inside a vacuum chamber. The material is vaporized from the target (in a plasma plume), which deposits it as a thin film on a substrate. The process is carried out in ultrahigh vacuum or in the presence of reactive gases like oxygen to deposit metal oxides. The process is shown in Figure8.8f [49].

b. **Chemical Vapor Deposition**
CVD is a heat-activated process and is based on the reaction of gaseous chemical compounds at the surface of substrate. The composition of gaseous compounds may vary from metal halides (chloride, bromide, iodide, or fluoride) or metal carbonyls, M(CO), as well as hydrides and organometallic compounds. The gaseous metal compounds are reduced or decomposed to the desired composition on a heated substrate. The coating deposition temperatures depending on chemical reactions vary from 800°C to 1200°C [54] (Figure 8.9).

The process variations in the CVD process include:

i. **Moderate temperature CVD – MTCVD**
The CVD process that takes place below 800°C is termed as MTCVD. The principle used is the application of reacting factor that requires lower activation energy. However, this process can be adopted for reactants like nitrogen and/or carbon donors [53].

ii. **Plasma-enhanced CVD:**
Using the electric plasma in the gas phase results in ionization and dissociation of chemical carrier gas and thus enhances chemical reaction and provides heat resulting in operation at lower temperature and improved deposition rate than MTCVD [54] (Figure 8.10).

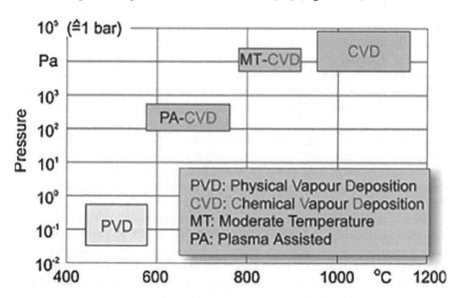

FIGURE 8.9   Variables of CVD techniques [54].

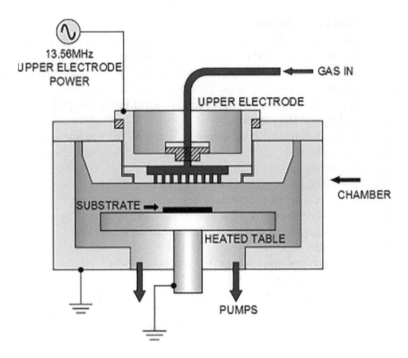

FIGURE 8.10   Chemical vapor deposition [54].

8.3.5 Thermal Spraying

As the name suggests, thermal spray-coating processes involve the deposition of coatings from a stream of high velocity finely distributed particles in a molten or semi-molten state impinging onto the substrate. The finely divided particles may be fine powder of metal wire. The continuous, directed, melt-spray consists of particles at 1–50 μm. The coating thickness varies from 0.5 to 2 mm. According to the heating sources to melt the particles, thermal spraying can be divided into flame, electrical arc, plasma, and detonation-gun spraying. This technique is capable of not only depositing metals but the range of materials varying from ceramics, polymers, and composites on varying types of substrates [55,56].

Depending on the energy sources used, processes are classified into three major categories: flame spray, electric arc spray, and plasma arc spray. The major advantages of this process include a wide variety of materials like metals, and that ceramics can be coated and the coating can be deposited without much heat input. However, the major disadvantage is that it is a line-of-sight process.

Thermal spray-coating methods typically involve rapid solidification with moderately high cooling rates of the order of 10^5–10^6 K/s [1]. Very high cooling rates result in the formation of nonequilibrium phases. Typically, the heterogenous microstructure consists of porosity, cracks, and impurities. Since the microstructure determines the final mechanical, thermal, and chemical properties of coating, extensive characterization of complex microstructure is mandatory [57].

The process may be classified as:

a. **Flame Spraying**

This includes the following:

i. Powder/wire flame

In this process, powder/wire is added into oxyfuel flame and is carried by flames and air jets onto the substrate. This is demonstrated by Figure 8.11a. Since the particle speed is relatively low resulting in poor cohesive strength, this leads to increased porosity and poor bond strength to substrate. The application of coatings is to recover worn machinery parts that can be mechanized later on [58,59].

ii. High-velocity oxy fuel (HVOF)

The heat source for flame spraying is a flame produced (as for gas welding) by the combustion of acetylene or propane in oxygen resulting in a temperature of 2500°C–3100°C, Figure 8.11b. The combustion takes place internally at a very high chamber pressure and exits through a small diameter of 8–9 mm. High temperature and velocity of injected particles result in dense and adherent coating having low porosity. This type of coatings can be used for high corrosion and wear-resistant applications where adhesion is also the main concern [57].

iii. Detonation gun spraying (D-gun spraying)

The equipment for D-gun spraying process is a long barrel that is closed at one end and opened at another end. The flammable gas mixture is composed of oxygen and fuel (most commonly acetylene), and the powder of material to be deposited is fed into a barrel as shown in Figure 8.11c.

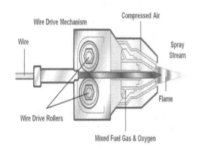

(a) Flame spraying [61]

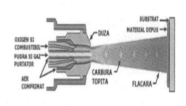

(b) HVOF [63]

(c) Detonation Gun [67]

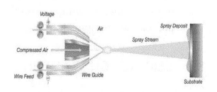

(d) Twin arc spraying [70]

FIGURE 8.11   (a) Flame spraying [70], (b) HVOF [60], (c) detonation gun [71], and (d) twin arc spraying [72].

The explosion is triggered by a sparkplug. This consequent intense combustion of the gas mixture leads to a detonation effect and the formation of high-pressure ultrasonic waves. The kinetic energy of the coating material particles is entirely transformed into heat energy, and a highly dense coating with strong adhesion can be deposited. The velocity of sprayed powders is about 800–1200 m/s; therefore, these types of coatings have low porosity and excellent adherence to the substrate. D-gun sprayed coatings are widely used in aviation, space flight, petroleum, metallurgy, and machinery industry [60–63].

b. **Electric Arc Spraying**

Arc spraying, also known as twin wire arc spray, uses an electric arc to melt wires. The atomization of molten metal is done to create the spray stream, which is then directed onto substrate to deposit coating, as depicted in Figure 8.11d. The process is easy to operate and automate. The typical feature of arc spraying is that the depositing material must be conductive and hence it is always a metal. The variation in the process includes simultaneous deposition of different materials/metals to produce a cladding, which is a mixture of both.

The process variables that affect droplet formation is dependent on the composition of metal melted and its behavior at high temperature (i.e., surface tension, dynamic viscosity, and heat conductivity), phenomena occurring during deposition (aerodynamic and magnetic forces), and wire configuration.

Because of the reaction of spray particles with compressed air, the deposits contain various oxides. The attractive feature of this process is low operating and capital cost of equipment, high energy efficiency, and high deposition rate due to which it is widely used in industries [64,65].

c. **Plasma Spraying**

Atmospheric plasma spray (APS), which involves generating a high-temperature plasma by passing a gas through a high voltage arc causing the molecules of the gas to split into ions and electrons, operates at power levels of 20–30 kW. The temperature reached is around 1000°C–2000°C. The most commonly used plasma gases are H_2, N_2, Ar, or He. The material to be added is fed into the arc, melted, and directed to the workpiece by plasma jet. The clad material may be in the form of powder or wire.

In addition to spraying of nonmetallic materials as cladding for metallic materials with electrically insulating layers, plasma spraying is one of the most common techniques to fabricate the high temperature insulating thermal barrier coatings [66–69].

8.4 SURFACE CHARACTERIZATION

In the case of various thermal, chemical, physical, and mechanical processes, such as oxidation, corrosion, adhesion, friction, wear, and erosion, surface plays a critical role. Hence, surface characterization techniques have established their importance in several scientific, industrial, and commercial fields and to understand the basic chemical and physical processes that affect material or device performance. By detailed understanding and analysis of the physical and chemical interactions that occur at the surface or at the interfaces of a material's layers, many of the problems associated with modern materials can be solved.

This section deals with the application of surface characterization techniques that can probe into complex surfaces and clarify their interactions in mechanical systems and processes.

8.4.1   X-RAY PHOTOELECTRON SPECTROSCOPY

X-ray photoelectron spectroscopy (XPS) is the tool for surface chemical compositional analysis. The surface chemistry influences factors such as corrosion rates, catalytic activity, adhesive properties, wettability, contact potential, and failure mechanisms.

XPS is a suitable method to investigate the composition of nanometer-thick surface oxide films. Both elemental and chemical state information of all elements (except He and H_2) can be obtained. The quantitative analysis of intensities of XPS measurements is useful to characterize the chemical composition, chemical states, and thickness of the passive films up to 3 nm. The accuracy of XPS is approximately $\pm 10\%$ [73].

8.4.1.1   Principle

The radiation used in XPS is monoenergetic soft aluminum Kα x-rays (1486.6 eV) or magnesium K α x-rays (1253.6 eV). Since the intensity of X-rays is small, the photon penetrates from 1 to 10 μm depth into a solid and emits the electrons by the photoelectric effect. The electrons originate from only the top few atomic layers (0.5–5 nm) because the mean-free paths of the electrons emitted are very small.

The kinetic energies of emitted electrons are given by $KE = h\nu - BE - W$, where hv is the energy of the photon, BE is the binding energy of the atomic orbital from which the electron originates, and W is the spectrometer work function.

The output of XPS analysis is given in terms of binding energy versus total electron count. Since the binding energy of electrons of different elements is different, it can be used to identify elements present. Chemical state (i.e., oxidation state, lattice sites, and molecular environments of the atom) plays an important role in the binding energy of the electron orbital and may change giving rise to observable energy shifts in the kinetic energy of the photoelectron giving information of chemical nature at the sample surface.

In many industrial and research applications, thin-film surface composition plays a very vital role in performance such as nanomaterials photovoltaics, catalysis, corrosion, adhesion, electronic devices and packaging, magnetic media, display technology, surface treatments, and thin-film coatings used for numerous applications [74–76].

8.4.2   ATOMIC FORCE MICROSCOPY

Atomic force microscopy (AFM) is a high-resolution imaging technique where a small probe with a sharp tip is scanned back and forth in a controlled manner, across a sample, to measure the surface topography at up to atomic resolution.

Nanoscale characterization of different material properties such as electrical, magnetic, and mechanical (modulus, stiffness, viscoelastic, and frictional) properties can be determined.

The AFM is capable of doing specifically three operations, as elaborated below:

Firstly, it can measure the force that exists between the probe and the specimen under it. Secondly, it can form a three-dimensional imaging profile as a result of the response of the probe on the specimen with respect to the force. This is accomplished by scanning the surface of the specimen. The third operation is controlling the force between the probe and the specimen in order to alter the properties of the sample.

The major advantages of AFM include easy sample preparation; can be operated in vacuum, liquid, or air; living system can be studied; and surface roughness can be quantified. However, limitation of the characterization includes limited vertical range and magnification [76–79].

8.4.3 NANOINDENTATION

As with conventional indentation technique, nanoindentation technique is not just a simple extension of microhardness test but is based on the theory of contact elasto-mechanics. Detailed analysis of loading and unloading curve is done to determine hardness, elastic modulus, viscoelasticity, creeping, fracture toughness, strain-hardening effect, residual stress, phase transition, and dislocation movement.

Nanoindentation method (also termed as depth-sensing indentation technique) gives the data about load displacement. According to the load–displacement data obtained from the tests, the elastic modulus and hardness can be derived from the slope of the initial portion of the unloading curve and the ratio of the peak load to the projected contact area of the indent, respectively [80–82] (Figure 8.12).

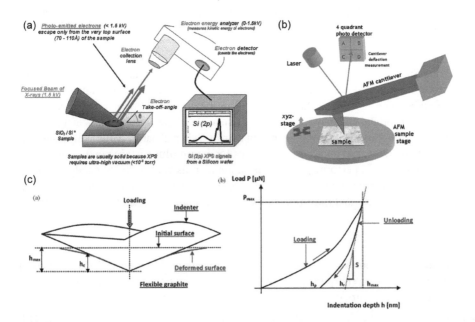

FIGURE 8.12 (a) XPS working principle [77], (b) atomic force microscopy [82], and (c) nanoindentation [83].

8.5 SUMMARY

Surface engineering plays major role in combating wear and corrosion, which accounts for a large proportion of GDP losses. Various surface engineering techniques have been discussed and explanation of advantages and disadvantages are summarized.

1. Microstructural modifications
   - Hardening process involves induction, flame, or laser having depth of 1–10 mm and possesses advantages like inexpensiveness and position specificity. However, this process has the disadvantage of distortion of samples.
   - Surface melting involves the formation of quasi liquid on surface. The major advantages of the process are ultrafine microstructure and high chemical homogeneity.
   - Shot peening involves plastic deformation of the surface resulting in increase in hardness, wear resistance, and fatigue resistance. However, the major drawback is that it modifies thin surface zone and results in a low increase in hardness.
2. Compositional modification
   - Electrodeposition deposits metal, oxide, or salt onto the surface of a conductor substrate by simple electrolysis of a solution containing the desired metal ion. The major advantages include high hardness and good corrosion resistance. However, the disadvantages include that the coatings cannot be deposited on complex geometries, and there is a risk of hydrogen embrittlement.
   - Hot dipping process includes dipping of sample in molten zinc and has advantages like good adhesion, good appearance, high hardness, and moderate wear resistance. The major drawbacks are the presence of porosity and difficulty to control coating thickness.
   - PVD process deposits dense coating with excellent adhesion and low coating process temperature, and pure metals, compounds, and alloys can be deposited. However, major drawbacks of the process include slow coating deposition of coating, high equipment cost, and line-of-sight process.
   - CVD process involves flowing a precursor gas/gas on one or more heated objects to be coated and chemical reactions occur on and near the hot surfaces, resulting in the deposition of a thin film on the surface. The major advantages are that a variety of materials can be deposited with very high purity, uniform coating thickness, very high hardness values, and good adhesion. However, major drawbacks are that the precursors used and gas produced are toxic and hazardous, only substrates that can sustain high temperatures can be used, large residual stresses are generated.
   - Thermal spraying utilizes combined thermal and kinetic energy of particles projected onto cleaned and ready surface where they adhere to

form a continuous coating. The coatings so produced have excellent adhesion and little heat input to substrate, and a wide variety of materials can be deposited. However, major drawbacks include line-of-sight process and over-spraying.

8.6 FUTURE WORK

This chapter gives an overview of the various techniques adopted for the deposition of metallic coatings. The future work may be directed toward the analysis and interpretation of recent techniques used for deposition of nanosized coating using cold spraying or RF/DC Magnetron sputtering technique or self-lubricating coatings including the graphene. Other latest development includes multifunctional organic-inorganic hybrids, nanocomposite materials (numerous silica- or/and siloxane, trivalent lanthanide ions, and other rare-earth metals), and films used in applications such as phosphors, lighting, integrated optics and optical telecommunications, solar cells, and biomedicine. Other developments consist of a combination of crystalline as well as an amorphous constituent; for example, TiN and Si_3N_4 used for high-speed machining or high-temperature metal-forming operations.

REFERENCES

1. J.R Davis. *Surface Engineering for Corrosion and Wear Resistance*. Materials Park, OH: The Materials Information Soeity, 2001.
2. K Holmberg, A Matthews. *Coatings Tribology: Properties, Techniques and Applications in Surface Engineering*. Tribology Series 28, Edited by D. Dowson. Amsterdam: Elsevier, 1994.
3. B Bunshah et al., eds. *Deposition Technologies for Films and Coatings: Developments and Applications*. Park Ridge, NJ: Noyes Publications, 1982.
4. T. Burakowski, T. Wierzchon. *Surface Engineering of Metals, principles, Equipment, Technologies*. Boca Raton, FL: CRC Press, 1998, pp. 22–23.
5. T. Burakowski, Surface Engineering yesterday, today and tomorrow, no 1, 1996, pp. 3–10
6. J. Dutta Majumdar, I. Manna. Laser surface engineering of titanium and its alloys for improved wear, corrosion and high-temperature oxidation resistance, In J. Lawrence, D. Waugh, Editors, *Laser Surface Engineering Processes and Applications*. Sawston: Woodhead Publishing, Series in Electronic and optical Materials, 2015, pp. 483–521.
7. V.V. Savinkin, P. Vizureanu, A.V. Sandu, T.Y. Ratushnaya, A.A. Ivanischev, A. Surleva. Improvement of the turbine blade surface phase structure recovered by plasma spraying. *Coatings* 10 (2020), 62.
8. H. Yao, Q.L. Zhang, F.Z. Kong. Laser remanufacturing to improve the erosion and corrosion resistance of metal components, In *Laser Surface Modification of Alloys for Corrosion and Erosion Resistance*, Edited by C.T. Kwok. Cambridge, UK: Woodhead Publishing Limited, pp. 320–355.
9. D. Kennedy, Y. Xue, M. Mihaylova. Current and future applications of surface engineering. *The Engineers Journal (Technical)* 59 (June, 2005), 287–292.
10. P. Sioshansi. Ion beam modification of materials for industry. *Thin Solid Films* 118 (1984), 61–71.
11. F. Malek Ghaini, M.H. Ameri, M.J. Torkamany. Surface transformation hardening of ductile cast iron by a 600w fiber laser. *Optik* 203 (February 2020), 163758.

12. S. So, H. Ki. Effect of specimen thickness on heat treatability in laser transformation hardening. *International Journal of Heat and Mass Transfer* 61 (June 2013), 266–276.
13. K.M.B. Karthikeyan, T. Balasubramanian, V. Thillaivanan, G. Vasanth Jangetti. Laser transformation hardening of EN24 alloy. *Steel Material Today Proceeding* 22(Part 4) (2020), 3048–3055.
14. Z. Li, J. Zhang, B. Dai, Y. Liu, Microstructure and corrosion resistance property of laser transformation hardening pre-hardened AISI P20 plastic die steel. *Optics & Laser Technology* 122 (February 2020), 105852.
15. Y.J. Cao, J.Q. Sun, F. Ma, Y.Y. Chen, X.Z. Cheng, X. Gao, K. Xie. Effect of the microstructure and residual stress on tribological behavior of induction hardened GCr15 steel. *Tribology International* 115 (November 2017), 108–115.
16. M.K. Lee, G.H. Kim, K.H. Kim, W.W. Kim. Effects of the surface temperature and cooling rate on the residual stresses in a flame hardening of 12Cr steel. *Journal of Materials Processing Technology* 176(1–3) (6 June 2006), 140–145.
17. T. Slatter, R. Lewis, A.H. Jones. The influence of induction hardening on the impact wear resistance of compacted graphite iron (CGI). *Wear* 270 (2011) 302–311.
18. https://www.laserline.com/en-int/hardening/.
19. https://www.tec-science.com/material-science/heat-treatment-steel/surface-hardening-case-hardening/.
20. Y. Li, S. Arthanaria, Y. Guan Influence of laser surface melting on the properties of MB26 and AZ80 magnesium alloys. *Surface and Coatings Technology* 378 (2019) 124964.
21. https://www.open.edu/openlearn/science-maths-technology/engineering-technology/manupedia/laser-surface-treatment.
22. K. Gupta, N. Kumar Jain, R. Laubscher. Chapter 6- Surface property enhancement of gears, In K. Gupta, N. Kumar Jain, R. Laubscher, Editors, *Advanced Gear Manufacturing and Finishing Classical and Modern Processes*, Academic Press, London, UK. 2017, pp. 167–196.
23. https://www.amtechinternational.com/shot-peening-benefits-gears/.
24. Q. Lin, H. Liu, C. Zhu, D. Chen, S. Zhou. Effects of different shot peening parameters on residual stress, surface roughness and cell size. *Surface and Coatings Technology* 398 (25 September 2020), 126054.
25. J. Walker, D.J. Thomas, Y. Gao. Effects of shot peening and pre-strain on the fatigue life of dual phase Martensitic and Bainitic steels. *Journal of Manufacturing Processes* 26 (April 2017), 419–424.
26. https://www.mfn.li/archive/issue_view.php?id=142.
27. A. Mallikand, B.C. Ray. Evolution of principle and practice of electrodeposited thin film: A review on effect of temperature and sonication. *International Journal of Electrochemistry* 2011 (2011), 1–16.
28. A. Gorokhovskya, M. Vikulova, J.I. Escalante-Garciac, E. Tretyachenko, I. Burmistrova, D. Kuznetsova, D. Yuri. Utilization of nickel-electroplating wastewaters in manufacturing of photocatalysts for water purification. *Process Safety and Environmental Protection* 134 (2020), 208–216.
29. H. Firouzi-Nerbin, F. Nasirpouri, E. Moslehifard. Pulse electrodeposition and corrosion properties of nanocrystalline nickel-chromium alloy coatings on copper substrate. *Journal of Alloys and Compounds* 822 (5 May 2020), 153712.
30. http://www.substech.com/dokuwiki/doku.php?id=electroplating.
31. X. Verdalet-Guardiola, B. Fori, J.-P. Bonino, S. Duluard, C. Blanc. Nucleation and growth mechanisms of trivalent chromium conversion coatings on 2024-T3 aluminium alloy. *Corrosion Science* 155 (2019), 109–112.
32. X. Liu, M. Wang, H. Li, L. Wang, Y. Xu Electrochemical effects of pH value on the corrosion inhibition and microstructure of cerium doped trivalent chromium conversion coating on Zn. *Corrosion Science* 167 (1 May 2020), 108538

33. J. Duszczyk, K. Siuzdak, T. Klimczuk, J. Strychalska-Nowak, A. Zaleska-Medynska. Manganese Phosphatizing Coatings: The Effects of Preparation Conditions on Surface Properties. *Materials* 11 (2018), 2585.
34. J Liu, B Zhang, W.H Qi, Y.G Deng, R.D.K Misra. Corrosion response of zinc phosphate conversion coating on steel fibres for concrete applications. *Journal of Materials Research and Technology* 9(3) (2020), 5912–5921.
35. B. Liu, X. Zhang, G.-Y. Xiao, Y.-P. Lu, B. Liu et al. Phosphate chemical conversion coatings on metallic substrates for biomedical application: A review. *Materials Science and Engineering C* 47 (2015), 97–104.
36. F. Jensen, I. Kongstad, K. Dirscherl, V. Chakravarthy Gudla, R. Ambat. High frequency pulse anodising of recycled 5006 aluminium alloy for optimised decorative appearance. *Surface and Coatings Technology* 368 (25 June 2019), 42–50.
37. V. Chakravarthy Gudla, K. Bordo, F. Jensen, S. Canulescu, S. Yuksel, A. Simar, R. Ambat. High frequency anodising of aluminium–TiO_2 surface composites: Anodising behaviour and optical appearance. *Surface and Coatings Technology* 277 (2015), 67–73
38. K.G. Swift, J.D. Booker. *Manufacturing Process Selection Handbook*. Amsterdam: Elsevier, 2013.
39. http://www.majesticanodising.com/about-anodising.php.
40. S.-P. Wang, L. Zhou, C.-J. Lia, Z.-X. Li, H.-Z. Li, L.-J. Yang. Morphology of composite coatings formed on Mo1 substrate using hot-dip aluminising and micro-arc oxidation techniques. *Applied Surface Science* 508 (2020), 144761.
41. https://galvanizeit.org/inspection-course/galvanizing-process.
42. https://en.wikipedia.org/wiki/Physical_vapor_deposition.
43. A. Baptista, F.J.G. Silva, J. Porteiro, J.L. Míguez, G. Pinto, L. Fernandes. On the physical vapour deposition (PVD): Evolution of magnetron sputtering processes for industrial applications. *Procedia Manufacturing* 17 (2018), 746–757.
44. A. Baptista, F. Silva, J. Porteiro, J. Míguez, G. Pinto. Sputtering physical vapour deposition (PVD) coatings: A critical review on process improvement and market trend demands. *Coatings* 8 (2018), 402
45. https://www.slideshare.net/muhammadaliasghar94/physical-vapor-deposition.
46. B.-K. Jang. Influence of temperature on thermophysical properties of EB-PVD porous coatings and dense ceramics of 4 mol% Y_2O_3-stabilized ZrO_2. *Journal of Alloys and Compounds* 480 (2009), 806–809.
47. F. Lomello, F. Sanchette, F. Schuster, M. Tabarant, A. Billard. Influence of bias voltage on properties of AlCrN coatings prepared by cathodic arc deposition. *Surface and Coatings Technology* 224 (2013), 77–81.
48. https://www.youtube.com/watch?v=JL8qvTW-WCg.
49. K.-D. Bouzakis, N. Michailidis, G. Skordaris, E. Bouzakis, D. Biermann, R. M'Saoubi. Cutting with coated tools: Coating technologies, characterization methods and performance optimization. *CIRP Annals - Manufacturing Technology* 61 (2012), 703–723
50. https://www.polifab.polimi.it/equipments/magnetron-sputtering-system-leybold-lh-z400/.
51. https://en.wikipedia.org/wiki/Electron-beam_physical_vapor_deposition.
52. R. Prabu, S. Ramesh, M. Savitha, M. Balachanda. Review of physical vapour deposition (PVD) Techniques. *ICSM Proceedings of the International Conference on "Sustainable Manufacturing"*, 2013, DoME & BIRC, Coimbatore Institute of Technology, Coimbatore.
53. F.-W. Bach, A. Laarmann, T. Wenz. *Modern Surface Technology*. Weinheim, Germany: WILEY-VCH Verlag GmbH & Co. KGaA, 2006, p. 80.
54. https://plasma.oxinst.com/campaigns/technology/chemical-vapour-deposition.
55. R. Younes, M.A. Bradai, A. Sadeddine, Y. Mouadji, A. Bilek, A. Benabbas. Effect of TiO_2 and ZrO_2 reinforcements on properties of Al_2O_3 coatings fabricated by thermal flame spraying. *Transactions of Nonferrous Metals Society of China* 26 (2016), 1345–1352.

56. H. Herman, S. Sampath, R. McCune. Thermal spray: Current status and future trends. *Thermal Spray Coatings* 25(7) (July 2000), 17–25.
57. M.M. Verdian, K. Raeissi, M. Salehi. Corrosion performance of HVOF and APS thermally sprayed NiTi intermetallic coatings in 3.5% NaCl solution. *Corrosion Science* 52 (2010), 1052–1059.
58. J.M. Guilemany, J. Fernandez, J. Delgado, A.V. Benedetti, F. Climent. Effects of thickness coating on the electrochemical behaviour of thermal spray Cr C –NiCr coatings. *Surface and Coatings Technology* 153 (2002), 107–113.
59. A. Belamri, A. Ati, M. Braccini, S. Azem. Hypereutectoid steel coatings obtained by thermal flame spraying — Effect of annealing on microstructure, tribological properties and adhesion energy. *Surface and Coatings Technology* 263 (2015), 86–99.
60. http://www.confind.ro/en/hvof.html.
61. H. Du, W. Hua, L. Jiangang. Influence of process variables on the qualities of detonation gun sprayed WC–Co coatings. *Materials Science and Engineering A* 408 (2005), 202–210.
62. H. Wu, X.-D. Lan, Y. Liu, F. Li, W.-D. Zhang, Z.-J. Chen, X.-F. Zai, H. Zeng. Fabrication, tribological and corrosion behaviors of detonation gun sprayed Fe-based metallic glass coating. *Transactions of Nonferrous Metals Society of China* 26 (2016), 1629–1637.
63. https://www.asminternational.org/documents/10192/1849770/06994g_chapter_sample. pdf.
64. P. Daram, P.R. Munroe, C. Banjongprasert. Microstructural evolution and nanoindentation of NiCrMoAl alloy coating deposited by arc spraying. *Surface and Coatings Technology* 391 (15 June 2020), 125565.
65. W. Tillmann, L. Hagen, D. Kokalj, M. Paulus, J. Metin Tolan. A study on the tribological behavior of vanadium-doped arc sprayed coatings. *Thermal Spray Technology* 26 (2017), 503–516.
66. A. Anupam, R. Sankar Kottada, S. Kashyap, A. Meghwal, B.S. Murty, C.C. Berndt, A.S.M. Ang. Understanding the microstructural evolution of high entropy alloy coatings manufactured by atmospheric plasma spray processing. DOI: 10.1016/j.apsusc.2019. 144117.
67. D. Francoa, H. Ageorgesa, E. Lópezb, F. Vargas. Tribological performance at high temperatures of alumina coatings applied by plasma spraying process onto a refractory material. *Surface and Coatings Technology* 371 (2019), 276–286.
68. M. Shia, Z. Xuea, Z. Zhang, X. Jid, E. Byonc, S. Zhang. Effect of spraying powder characteristics on mechanical and thermal shock properties of plasma-sprayed YSZ thermal barrier coating. *Surface and Coatings Technology* 395 (2020), 125913.
69. https://www.sciencelearn.org.nz/images/250-plasma-spray-process.
70. https://www.metallisation.com/product category/flamespray/.
71. https://www.plakart.pro/en/technologies/detonation-spraying-d-gun/.
72. https://www.oerlikon.com/metco/en/products-services/coating-equipment/thermal-spray/processes/electric-arc-wire/.
73. J.E. deVries. Surface characterization methods—XPS, TOF-SIMS, and SAM a complimentary ensemble of tools. *Journal of Materials Engineering and Performance*, 7(3) (1998), 303–311.
74. M. Aziz, A.F. Ismail. Chapter 5- X-ray photoelectron spectroscopy (XPS), In N. Hilal, A. Ismail, T. Matsuura, D. Oatley-Radcliffe, Editors, *Membrane Characterization*, Elsevier Science, London, UK. 2017, pp. 81–93.
75. V. Maurice, W.P. Yang, P. Marcus. X-ray photoelectron spectroscopy and scanning tunneling microscopy study of passive films formed on (100) Fe—i 8Cr—i 3Ni single-crystal surfaces. *Journal of the Electrochemical Society*, 145(3) (March 1998) 909.
76. K. Miyoshi. *Surface Characterization Techniques: An Overview.* Cleveland, OH: Kazuhisa Miyoshi Glenn Research Center.

77. https://en.wikipedia.org/wiki/X-ray_photoelectron_spectroscopy.

78. R.R.L. De Oliveira, D.A.C. Albuquerque, T.G.S. Cruz, F.M. Yamaji, F.L. Leite. Measurement of the nanoscale roughness by atomic force microscopy: Basic principles and applications. *Atomic Force Microscopy - Imaging, Measuring and Manipulating Surfaces at the Atomic Scale*, (March 23rd 2012) 147–174.

79. M.K. Khan, Q.Y. Wang, M.E. Fitzpatrick. Atomic force microscopy (AFM) for materials characterization, In G. Hübschen, I. Altpeter, R. Tschuncky, H-G. Herrmann, Editors, *Materials Characterization Using Nondestructive Evaluation (NDE) Methods*, Woodhead Publishing, London, UK. 2016, pp. 1–16.

80. Wikipedia. Atomic force microscopy. https://en.wikipedia.org/wiki/Atomic-force_microscopy. Assessed 7 June 2016.

81. W. Zhao, W. Song, L.-Z. Cheong, D. Wang, H. Li, B. Flemming, F. Huang, C. Shen. Beyond imaging: Applications of atomic force microscopy for the study of Lithium-ion batteries. *Ultramicroscopy* 204 (2019), 34–48

82. https://simple.wikipedia.org/wiki/Atomic_force_microscope.

83. M. Khelifa, V. Fierro, J. Macutkevic, A. Celzard. Nanoindentation of flexible graphite: Experimental versus simulation studies. *Advanced Material Science* 3 (2018): DOI: 10.15761/AMS.1000142.

9 Surface Coatings for Automotive Applications

Ashok Raj J.
Bharath Institute of Higher Education

Santosh Kumar B. H.
GITAM School of Technology

L. Arulmani
R R Institute of Technology

Jitendra Kumar Katiyar
SRM Institute of Science and Technology

CONTENTS

9.1 BACKGROUND AND INTRODUCTION

Tribology is the study of contacting surfaces which relatively move each other. It can be interpreted as dealing with the phenomena associated with friction, wear and lubrication [1]. Erosion and wear directly affect machinery's lifetime and power consumption. Erosion and wear lead to loss of material from the moving parts of the machinery and affect safe working. Hence, it becomes necessary to control or reduce erosion and wear of the surface [1–3].

DOI: 10.1201/9781003097082-9

In recent years, the abovementioned objective has been accomplished by making changes in the design, using materials with improved bulk or with the application of better lubrication technology. Solid lubricants or synthetic oils or mineral oils are used for better lubrication and ceramics or polymers as bulk materials [1–4].

In search of more approaches towards reduction in friction and wear control, tribologists are utilizing surface treatments and coatings that have opened the new field of surface engineering. Two main factors are involved in this growth. The development in new coating technology and treatment method is the first reason. This development has provided coating characteristics and tribo-chemical properties, which were never achieved before. The second reason surface behaviour is observed by material scientists and engineers because behaviour of surface is crucial in many engineering mechanisms [2–4].

Surface tends to be the origin for most of the failures that could be caused either by corrosion, wear or fatigue. The surface is affected by the electronics, optical and magnetic characteristics apart from mechanical and chemical properties [3]. Improved technologies, automation in various fields and the increased rate of production have made high-performance mechanical components and tools crucial. Desired surface properties in machine components can be achieved through surface engineering. Like coating over the surface, it can be used for improved wear resistance, corrosion resistance and thermal load resistance [4].

Even before fully optimized coatings were available, tribologists provided analytical fundamentals for the prognosis of suitable parameters of surfaces that were developed by sheer grasp of the surfaces behaviour in contact. This became the platform for the development of new coatings and treatment methods and has created a significant impact [1–2]. Operation of devices and bearing systems working at near-vacuum conditions or machine components working beneath highly corrosive and erosive environments are possible with the support of advanced tribological coatings [1].

Depending on the state of the deposition phase, the coating process can be classified into four generic categories, namely, solid, molten, solution and gaseous [1–5].

As per the tribological considerations and requirements on counter-surface using surface coatings [2–4] are as follows:

a. Friction coefficient is either initial or steady-state and its instability must be within permissible deign limit.
b. The wear on the coated surface and its counter surface must not surpass a critical design limit.
c. The life of any system must be longer than the predicted life within specified conditions. The limit of any system life might be defined as happening even though one of the previous conditions is not maintained.

The coatings must exhibit a combination of certain properties, such as hardness, thermal expansion, shear strength, elasticity, fracture toughness and adhesion to meet tribological requirements. Figure 9.1 represents the difference between various zones of the coated surface. Each zone consists of different properties, which is necessary to consider [5].

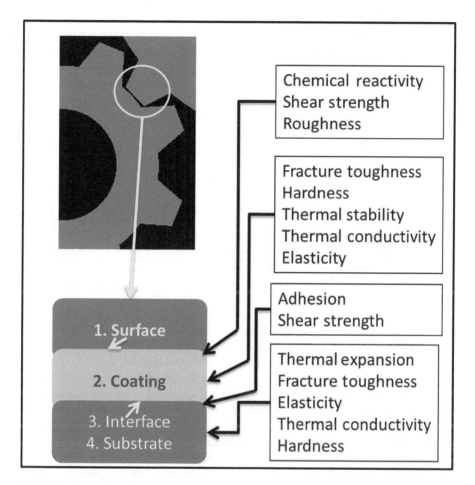

FIGURE 9.1 The properties of various zones on the coated surface.

The coating and the substrate must possess certain important properties such as strength of materials and their thermal characteristics, which are investigated by their microstructure, porosity, homogeneity and composition. The shear strength and adhesion of the counter-surface at the interface between substrate and coating play an important role in their tribological characteristics. Apart from shear strength, properties such as chemical reactivity and the roughness at the coating surface must be considered [6–8].

Surface design is not easy as there are some fundamental problems to achieve all the required properties. Desired properties such as no surface interactions with the counter-faces and better adhesion at the coating–substrate interface or high toughness and high hardness cannot be obtained at the same time. For instance, toughness and adherence decrease with the increase in hardness and strength. Hence, an optimum way is required between the properties of surface coatings and economic concerns [7].

The coating process and its thickness contribute majorly towards in selection of material for coating and its fabrication parameters as shown in Figure 9.2.

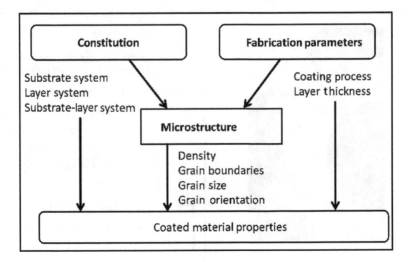

FIGURE 9.2   The factor influenced on material properties of surface coating [22].

The microstructure, density, grain boundary, grain size and grain of the coating depend on the coating process and thickness. To gain maximum benefits, functional parameters and properties of the coating surface can be optimized.

The parameters that make the benchmark for the selection of engineered surface over another surface are often complicated because coating or surface treatment can change the properties of the bulk material. Hence, to obtain greater performance through coating, it is important to redesign the procedure. The availability of many coatings has put the design engineers in dilemma in selecting coatings [4–7].

Different thoughts have been given on the selection procedure for surface coatings. Many design engineers would adopt the selection approach. Selecting better coating by rejecting those with poor results [5].

The surface plays a very crucial role in the part production with better performance at a feasible price, and this is possible if we have the freedom to take advice and suggestion before finalizing the design. Hence, it's necessary to take expert advice available at the earliest prospect. Tribological coatings provide improved quality and performance by using lower grade and low-cost substrate materials [1–5].

Recent advancements in computer simulations have contributed significantly to coating selection process by reducing efforts and cost. Simulation makes feasible the implementation of practical solutions that could earlier only be achieved in theory. In many tribological systems, tribochemical reactions occurring at the contact interface play a critical role. The compounds formed from the reaction contribute and control the tribological behaviour at interface. The formation of specific compounds can help improve the wear resistance. This compound can be achieved by using extreme pressure additives. Therefore, advanced techniques of surface engineering tolerate the design and production of such compounds over surface as a layer [5–8].

The above possibilities not only result in improved sliding mechanism but also enhanced the thermal behaviour. For this reason, both thermal diffusion and chemical diffusion are influences. As a consequence, there is a possibility of the efficiency of machines being improved with the substantial economic advantages of a good tribological design. An illustrative example is the estimated potential improvement in the efficiency of pumps, which in their various forms are reputed to consume roughly 10% of the total energy consumption in industrialized countries and up to 20% in process industries. Friction and wear can be reduced by using improved coatings and a better selection procedure. Increased friction and wear contribute towards reduced efficiency of the machine by increasing overall energy consumption and increase cost by increased in-service failure charge or maintenance charge. Hence, it is crucial to select improved coating and selection procedures for reduced friction and wear. It follows that a major goal for surface coatings is that it should improve the extended and predicted life of components and behaves ideally in non-catastrophic failure modes [5–8].

To meet the demand, market competition and face challenges, design engineers, researchers and scientists are exploring new materials and technology. Several studies have been conducted to improve the friction and wear properties, components' working life, efficiency of fuel, corrosion resistance, surface morphology and to reduce weight and emission. In an IC engine, the friction and wear performance of the piston ring has long been established as an important parameter for obtaining a required engine efficiency and durability such as power losses, consumption of fuel and oil, blow by, and even destructive emissions through exhaust [6–8].

There is considerable loading of physicochemical degradation to the cutting tools surface and dies in the manufacturing environments of automotive industries that make it quite complicated. Wear is a phenomenon described by how a tool will last in service. The tool involvements superior loads, high speeds and friction and as a consequence elevated temperature. These factors are responsible for wear of tool surfaces. The failure of tool surfaces is curtailed by lubrication. The liquid lubricants fail under extreme conditions and are moreover environmentally inadmissible. Surface coatings can improve the tool surface, improves the life of tool and enhances the quality of manufactured components [1–4].

Surface design and characteristics play a crucial role along with the bulk material properties in automotive components like engine and power train components. So in these components and its surface must achieve engineering functions in various critical conditions and hence shall require treatment to improve its characteristics. In addition to wear protection, surface coatings in automotive components can improve corrosion resistance; enhance surface hardness and reduce weight. The other benefits of surface coatings in automotive components include reduced oil and fuel consumption, reduced CO_2 emissions and extended durability [5].

9.2 SURFACE COATINGS FOR CUTTING TOOLS

Modern machining processes face repeated cost pressure, productivity issues and high-quality potentials. Before the development of surface coating technology, the use of cutting fluids was considered an essential part of the machining process.

During the machining process, due to the relative motion of workpiece and tool, the frictional heat is generated because of chip–tool interaction. To reduce generated temperature and control friction in the cutting zone, cutting fluid was used earlier [1–3]. Cutting fluid also serves the purpose of washing away chips and prevention of adhesion between tool and workpiece. For a stable cutting process and to obtain good surface finish with improved tool life, cutting fluids are crucial. However, cutting fluids come with a few disadvantages such as formation of toxic vapours, smoke fumes and bad odour creating unpleasant working environment. Sometimes, the formation of bacteria can affect operation. The undesirable effects of using cutting fluid were considered to be necessary evil [4–5].

Surface-coated cutting tools that possess high hot hardness are an important development in dry machining. Nodular cast iron can be used in automotive applications such as crankshaft, piston, gears, flywheel, dies, etc. The machinability of nodular cast iron is less than that of grey cast iron. During its machining, the surfaces of the cutting tool need to be hard, wear-resistant and chemically stable. Surface engineered coated tools can be used to accomplish this objective. Yigit compared the effect of coated and uncoated tools on turning of nodular cast iron. Uncoated tools performed badly in terms of tool wear and surface finish. It was found that at higher cutting speed, a 10.5 μm multilayer TiCN/TiC/TiCN/Al$_2$O$_3$/TiN coated carbide tool is the most suitable for nodular cast iron turning. Moreover, diamond-coated cutting tools were used for machining of nonferrous materials such as aluminium alloys, copper alloys, fibre reinforced polymers, green ceramics and graphite, etc. These tools are extremely hard, possess high tool life and ensure precision machining. However, the adherence of diamond film on substrates, surface roughness of films and the reduction in the transverse rupture strength of coated substrates must be explained for the widespread use of cutting tools coated by diamond [1–5].

9.3   SURFACE COATING MATERIALS AND TECHNIQUES

Four types of coating materials are used by industries to develop hard and wear-resistant surfaces on cutting tools. The first type includes titanium-based coating materials such as TiN, TiC and TiCN. The second type includes ceramic coatings like Al$_2$O$_3$. The third type represents super hard coatings like polycrystalline cubic boron nitride (PCBN) and chemical vapour disposition diamond. The fourth type includes coatings that have self-lubricating properties with lower friction coefficient such as amorphous metal carbon. Physical vapor deposition (PVD) is a method of coating under vacuum conditions. In this process, the coating material is removed physically by the application of heat due to which evaporation or sputtering occurs. These evaporated particles condensed on the surface as a layer of thin film. The temperature range of this process is kept between 450°C and 550°C for film deposition on high-speed steel substrates. Hence, PVD coatings control thickness on edges leading to a sharp-coated edge. The ability to de-coat and re-sharpen the PVD-coated tools makes it suitable for automotive industries. Further, chemical vapour deposition (CVD) is another method to deposit the thin film on substrate through the chemical reaction of gaseous compounds with preheated substrates. It has the ability to produce thick layers by CVD with higher deposition rates makes it highly suitable for a high metal removal rate. Owing to the interaction of precursor gases with substrates in CVD Coatings, it may be generated the brittle carbides sometimes [5–7].

9.4 METHODS OF COATING THE SURFACE

In the present era, because of advanced methods for coatings, there has been an accelerated development of tribological coatings that cater to the attributes previously unachievable. Properties that were difficult to achieve were as follows: morphology, strength, composition and adhesion [1]. The plasma and ion-based methods for deposition techniques have been drastically studied and used. Therefore, this chapter describes various processes used for surface coatings

Large-scale commercial exploitation of plasma and ion-based processes were delayed to date mainly due to the following difficulties with advanced technologies. These are high voltage and current technology, process control-related electronic technologies, the physics and chemistry of plasma and coating under vacuum. Plasma-based techniques, however, offer considerable benefits to various sectors of engineering, providing us the much-needed solutions required to advance in this field. It is better to not neglect other methods as the traditional processes with significant improvement have responded to the challenges of the new techniques and methods. Hence, this chapter describes briefly various coating methods. These processes are largely divided into four categories such as solid state, molten or semi-molten state, solution state and gaseous state processes, respectively.

Solid-state processes tend to produce relatively thick coatings. Solid-state coatings will not be studied in this chapter as we are concerned with thin coatings. The general classification of surface engineering technique is shown in Figure 9.3. The most crucial process constraints for coatings are their thickness and the deposition temperature.

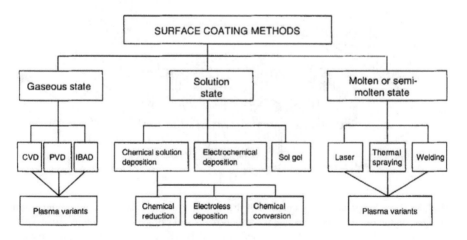

FIGURE 9.3 Flow chart for surface coating methods.

The thickness of coating typically ranges between 0.1 and 10 mm. Further, the temperature varies from room temperature up to 1000°C for deposition over the surface which is shown in Figure 9.4. This is the development of newer methods that acronyms the life.

9.4.1 GASEOUS STATE PROCESSES

In this method, the targeted material is vaporized in vacuum by providing heating. That vaporized material coated over the targeted surfaces for modification of surface. There are two generally used methods, i.e., CVD and PVD. In the CVD method, the precursor gas is used as a reagent source through which coating over the surface occurs whereas, in the PVD method, the coating materials are evaporated or atomized from solid state to gaseous state over the targeted surface. Both methods have their importance in surface engineering because of several reasons. Using these methods, pure ceramic films can also be deposited over any surface. Furthermore, in the PVD method, the energy of coating material is increased up to ionized state, and then these ions accelerate towards the development of film over substrate. This can also be obtained by using ion beam source or by utilizing plasma around the surface through which ion can be accelerated [10]. The advantages of these plasma-assisted processes are listed below:

a. Coating adhesion is improved due to the potential to preheat and clean the surface using ion energy and neutral bombardment. This method is known as cleaning, due to sputtering,

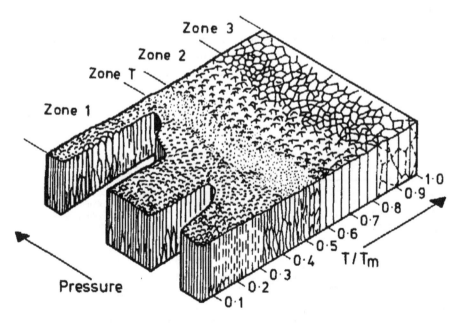

FIGURE 9.4 Schematic representation of the influence of substrate temperature and argon pressure on the microstructure of metal coatings deposited using cylindrical magnetron sputtering sources. T (K) is the substrate temperature and Tm (K) is the melting point of the coating material [23].

b. The thickness of the coating can be maintained uniformly through gas-scattering effects. During the deposition process, the sample can be displaced or rotated relative to the vapour source. This helps in obtaining uniform coating thickness.

c. In most of the cases, the coating produces the final surface finish. Hence, after coating, no additional machining or polishing is required.

d. Due to bombardment, columnar growth obtained which encouraged the atom transfer, causing the controlled coating structure.

e. A large variety of materials can be coated, including insulators.

f. Vapour source such as resistance heater, electron beam and sputter magnetron help in controlling deposition rate.

g. Most of the processes produce effluents or pollutants. In this process, no such pollutants are produced as hazardous by-products.

h. Controlled vacuum environment and pure source materials help in high purity deposits.

Figure 9.4 shows the resilience of the PVD coating process. It shows that how the structure of coating varied as a function of deposition temperature and pressure, according to Thornton (1974).

9.4.1.1 Applications of CVD Coatings

This coating method has great importance in industrial applications, which are as follows:

- Wear resistance, erosion protection coatings, high-temperature protection and corrosion resistance.
- CVD is useful in dense structural parts that are difficult or expensive to produce using traditional methods or techniques. Optical Fibres – For telecommunications.
- Nano-machines.
- Composites – Preforms can be infiltrated using CVD techniques to produce ceramic matrix composites such as carbon-carbon, carbon-silicon carbide and silicon carbide-silicon carbide composites. This process is sometimes called chemical vapour infiltration or CVI.
- Powder production – Production of novel powders and fibres.
- Devices such as optoelectronic, sensors and integrated circuits, made using semiconductors.
- Used as catalysts.

9.4.1.2 Applications of PVD Coatings

PVD coatings, in general, are used to improve hardness, oxidation resistance and wear resistance. Hence, their applications are in a wide range such as:

PVD coatings can be used to coat any material on anything on anything. PVD-based deposition techniques have the following advantages.

1. High rate of material deposition
2. Ease of sputtering any metal, alloy or compound

3. Deposition films are highly pure
4. Films possess extremely high adhesion
5. Excellent coverage of steps and small features
6. Heat sensitive substrates can be coated using this technique
7. Ease of automation
8. Uniformity obtained on large-area substrates is excellent

9.4.2 Molten and Semi-Molten State Processes

In this method, laser surface treatment and hard facing techniques such as thermal spray and welding are included. This method changes the physical and mechanical properties of the substrate surface. A brief description of these methods is described≈here.

9.4.2.1 Laser Surface Treatments

In recent years, laser surface treatment has drastically grown in the field of laser utilization. The treatment through laser on surface can produce a greater range of important surface properties. Laser surface treatment allows the use of cheap base materials. It can be applied to enhance the fatigue life of machine components due to the hard surface layer and decrease the wear over surface. This method work on the principle of beam of coherent light source fall on the surface with high intensity of power due to which surface is modified. Laser surface treatment can be used in vacuum, protective or progressive gases.

An optical transmission system is used to direct light onto the surface. Optical instruments such as lenses, scanner units, mirrors and beam integrators are used to modify beam focusing, thereby controlling the mean optical output power and power density.

Track patter can be consecutively produced over the surface of a component through the movement of the laser beam. The feed rate and cross-section of the substrate surface determine interaction time [5–9].

9.4.2.1.1 *Importance and Limitation of Laser Surface Treatment*
The importance of laser surface treatment can be studied from various aspects:

Inadequate energy depletion and its significances:
Less amount of energy is required in this process as compared to the conventional surface heat treatment process. Due to the localized heating, there is minimal energy input, minimal deformation or dimensional changes. Furthermore, minimal machining of component is required, which also saves energy [1,3,18].

Environmental aspects:
The surface is conducted to the cold bulk of the material producing a hardened surface. Self-quenching helps in the formation of hardened surfaces by generation of heat. There is no requirement of cleaning and washing the workpiece after the heat treatment as the process is carried out by any quenching agents [5–11].

Flexibility and productivity of the process:

Different degrees of defocus make it possible to control energy input. Laser source power can be easily changed by different focuses using focusing lenses. Competence of switching the laser beam between workstations by simple optical devices in the atmospheric environment [5,12,19,20].

The use of various shapes of lenses and mirror makes it easy to work on different shapes and complexity of the product. Heat treatment of very large or small parts with complex shapes can be done using this method. This method is suitable for both individual and mass production. Reproducibility and reliability of quality of the treated surface layer with accurately controlled depth and width of the layer; high level of suitability for production line incorporation and automation of the procedure [2,7,9,21].

9.4.2.1.2 Use of Laser for Heat Treatment Can Have the Following Difficulties

Non-homogeneous energy distribution in the laser beam; very narrow temperature field ensuring the aimed micro-structural changes; adjustment of kinematic conditions of the workpiece and/or laser beam to different product shapes; and reduced absorptivity of the laser light in interacting with the metal material surface [5,9,12].

As with all other power beam methods, a treated surface must progressive built-up by a series of overlapping tracks. Track widths differ according to the type of laser interaction time and power density but are typically ~0.5–2 mm. Hence, numerous power beam passes are needed before a treatment is completed. However, because power beam scan speed is frequently rapid, economic treatment times per unit component are still possible [8,16,21].

9.4.2.2 Thermal-Sprayed Coatings

Thermal-spraying technology makes use of combustion flame or electrical energy to fuse a wide range of materials. Thermal spraying has the potential to produce high-quality and high-performing surface coverings. Thermal spraying is very flexible as it can be for a wide variety of types and shapes with film thickness varying from thick to thin. This technology is heavily used in all types of industrial fields for upgrading and diversification, as well as improving substrates performance, and development of substrates with new functions [3–5].

9.4.2.2.1 Features of Thermal Spray Technology

Thermal spray technology has a variety of superior features when compared to other surface covering technologies [1–5].

- Depending on the purpose of modification and application, a wide variety of materials can be used as thermal-spraying materials. Spraying materials such as fusible alumina, ceramics, as well as metals and plastics are used.
- From micron to mm thickness, small to large parts and complex-shaped equipment can be modified.
- The organizational and physical properties of the subject remain unchanged as the thermal spray substrate remains comparatively low temperature during spraying.

- Depending on the modification purpose and intended application, thermal spraying can be carried out in atmospheric conditions, low-vacuum, or inert gas environments.

9.5 RECENT ADVANCES IN SURFACE COATINGS FOR AUTOMOTIVE APPLICATIONS

In recent years, we have seen drastic advancements in the PVD technology. The PVD technology has wide application in the automotive industry. It is the key contributor to its development, where hard coatings have an efficient tribological function. Apart from acting as a heat carrier from the piston to the cold cylinder, the major role of the piston ring is to seal off the majority of combustion gas, which directly adds to the power output of the engine. Tribological performance is one of the most important parameters in designing piston rings for heavy-duty diesel engines. Piston rings are under high mechanical loading due to motion and high temperature leads to softening of the ring material leading to a high rate of piston ring wear during operation. Hence, for the efficient performance of engines, it is mandatory to reduce wear rate, contributing to reliability over a long lifetime. In this regard, several investigations proposing metal nitride coatings for tribological (piston ring's functional surface) purposes have been published [1–3]. CrN, in comparison with other conventional metal nitride coatings, has shown excellent results towards corrosion resistance in both environmental and high-temperature conditions. Even a high rate of deposition for similar deposition parameters is observed.

Cost efficiency with the added advantage of high wear resistance and scuffing resistance is the reason for considering CrN as coating materials for piston rings in large-scale production industries even these days. Regarding CrN coatings development and deposition, the analysis of the influence of the deposition conditions in the final coating properties or the study of the CrN coating tribological response were study aims of some of those investigations [4–9].

Friedrich et al. [4] studied Cr_xN coatings for tribological applications on piston rings. From the investigation, it was observed that, in comparison with Cr coatings, all coatings deposited with a reactive nitrogen flow showed superior hardness. The same hardness value of 30GPa was observed for coatings of $15\,cm^3/min$ (Cr_2N) and $25\,cm^3/min$ (CrN) of nitrogen gas flow. Owing to the reduction in the plasticity of the contact, the friction coefficient is slightly decreased with increasing hardness was another key observation of this study.

Broszeit et al. [5] continued working on the development of CrxN coatings. In this investigation, the tribological performance of CrN coatings deposited by PVD and electroplated chromium coatings was compared, and the former reported better results. In other studies on CrN coatings, Barata et al. [6] worked on the influence of magnetron sputtering deposition parameters in the crystal phase, and Bouzid et al. [8] conducted studies on CrN deposited by RF magnetron sputtering and investigated tribological resistance. In the latter work, he found that the non-alloyed CrN layer CrN coating achieved the lowest coefficient of friction.

The deposit rate of magnetron sputtering compared to other PVD techniques is low, making it not feasible for the automotive industry where the requirement is large-scale production. This has become a major drawback of magnetron sputtering technology. Different attempts are made to enhance the deposition rate of the magnetron sputtering technique using different variants. High Power Impulse Magnetron Sputtering (HiPIMS), a recent advancement technique, focused on the production of highly ionized fluxes of the sputtered material. However, the cathodic arc technique is one of the PVD techniques employed for improved deposition efficiency (regarding the productiveness and the financial feasibility) [10]. Owing to improved deposition rate, the cathode arc technique fulfils the requirement of the large-scale production industry. For this reason, the cathode arc technique has been widely used for protective and tribological coatings in several industries, such as the automotive one.

A brief overview of this process can be used to describe the coating deposition in three steps. When a low-voltage and high-current arc is struck between a conducting cathode and an electrode, vaporization of the solid cathode material occurs through the arc erosion. Next, gaseous plasma is formed from the vaporized cathode material, which acts as a fluid conductor between anode and cathode (electrodes). In the last step, plasma ions condensation occurs and arrives at the substrate surface [11]. It was found that the arc discharge and plasma density produced are directly proportional [12].

Cathode material undergoes four phases in the cathode spot region due to the presence of high energy density. The phase change is from solid to plasma with possible liquid and intermediate phases. Three different types of cathode erosion: irons, neutrals and macroparticles occur as a result of plasma transformation. The difference in plasma potential and substance surface potential gives the energy of ions. The energy of condensing plasma is affected by substrate surface potential, which can be controlled by the applied bias voltage. The final properties of the deposited coating can be controlled by varying different deposition parameters using the cathode arc technique.

Improved processing and paint chemistries have led to the noval development in paint pigments. For example, depending on the angle at which they are viewed, flake-based pigments based on aluminium and interference pigments change their colour (otherwise known as the "flip flopping" effect) has given customer satisfaction of automotive coatings by enhancing brilliance, colour and appearance [11]. New spray guns and spray gun configurations have been developed to handle the difficulty of using these new pigments in spray gun technology. Computer-controlled spray guns drastically reduced the need for spray painting craftsmanship. Earlier painting was done manually.

Further, worker safety and increase in the ratio of deposited paint-to-paint sprayed are improved due to automation. The most expensive operational aspect and a major energy-consuming area of an automobile assembly is the automobile paint shop. It consumes 30%–50% of the total costs of the manufacturing of automobiles [13]. A major part of this cost can be added to the cost of energy used for air handling and conditioning (HVAC) as well as for treatment of emission generated and paint drying, paint booths should be designed and must be capable enough to remove overspray paint particles, evaporated solvent and any other related pollutants like VOCs. Hence, the energy associated with only booth ventilation is significant [14]. In general, painting operations contribute to 70% of the total energy costs in assembly plants [15].

Energy used to dry 200-μm film on an automobile surface is not significant, but it is important to consider all activities to complete painting and their cost, such as heating of paint, the carriers, on which the body of an automobile is carried, underlying automobile body and the dollies. Because of the benefits of inorganic pretreatment, liquid base coats, liquid or powder primer surfaces, cathodic electro-deposition, and one- or two-component solvent-borne clear coats, automobile painting processes are more standardized these days. For example, development in powder coatings and additional advantage of meeting environmental regulations have made use of it in many car manufacturers. The reliability of powdered coatings is high, hence, most of the manufacturers have decided to use them [16].

In North America at Chrysler in all running manufacturing units, at GM for their truck plants, and in all new paint shops, powder coatings are now used throughout all primer surface operations. Powder coatings have been used for the clearing of the coat process, in some plants, in Europe [17]. This increase of powder coating applications has coincided with a drastic shift in the type of materials used in automobile body construction. Formerly made mostly of steel, today's automobile bodies are made up of 30% of aluminium and high-strength steel. Other lightweight materials are also being applied, including magnesium and polymer composites made of glass and carbon-fibre-reinforced thermosets and thermoplastics [18]. Over a period of time, automotive coatings have evolved as they either satisfy or are anticipated to meet customer expectations and environmental regulations while also lowering manufacturing and ownership costs.

Smart coatings significantly improve surface durability. Additional properties such as self-healing, soundproofing, self-stratifying, self-sensing, super-hydrophobicity and vibration damping are achieved from smart coating. For instance, a smart coating could use to enhance the coating life in response to the environment; to reduce abrasion, smart coatings can be used as self-healer, mechanical trigger or to a corrosive event in which the coating is self-healing as a result of UV, heat, or mechanical activation [19]. Self-healing in coating can also be achieved with the use of shape memory polymers that are triggered with temperature, UV radiation and humidity manipulations; self-healing related to the swelling of special clays such as montmorillonite is also possible [20]. Coatings with internal sensing capabilities that entail the passive or active triggering of fluorescent molecules or quantum dots fall under the category of smart coatings [21]. In the former, the sensing system signals and activates changes in or repair of the coating by sending data to an external detector; in the latter, the sensing system itself would be responsible for outputting the response signal.

9.6 CONCLUSIONS AND FUTURE WORKS

In this chapter, this important point about coatings has been observed: it exhibits a combination of certain properties, such as hardness, thermal expansion, shear strength, elasticity, fracture toughness and adhesion to meet the tribological requirements. Further, the coating process and the thickness of the coating contribute majorly towards the selection of coating material and the fabrication parameters. To gain maximum benefits, functional parameters and properties of the coating surface can be optimized. Moreover, surface design and characteristics play a crucial role along with the bulk material properties in automotive components like engine and power train components. Surface coatings in automotive components can improve corrosion resistance; enhance surface hardness and weight reduction. The other benefits of surface coatings in automotive components include reduced oil and fuel consumption, reduced CO_2 emissions and extended durability. There are various methods being employed for surface coatings. These are gaseous state processes, solution-state processes, molten or semi-molten state processes and solid-state processes. Among all coating processes, the PVD technology has wide application in the automotive industry. It is the key contributor to its development, where hard coatings have an efficient tribological function. Further, self-healing in coating can also be achieved with the use of shape memory polymers.

In the future, the development of a new methodology for surface coating will be adopted such as surface texturing followed by surface coating, self-healing coatings, self-lubricating coatings, etc. The hybrid coatings combination will be also play a very important role in the near future.

REFERENCES

1. M.F. Ashby, D.R.H. Jones, *Engineering Materials: 2. An Introduction to Microstructures, Processing and Design*, Pergamon Press, Oxford, 1986, p. 4.
2. H. Skulev, S. Malinov, P.A.M. Basheer, W. Sha, *Surface and Coatings Technology*, 185(1), 1 July 2004, 18–29. DOI: 10.1016/j.surfcoat.2003.12.012.
3. L.M. Berger, Application of hard metals as thermal spray coatings, *International Journal of Refractory Metals and Hard Materials*, 49, March 2015, 350–364. DOI: 10.1016/j.ijrmhm.2014.09.029.
4. C. Friedrich, G. Berg, E. Broszeit, F. Rick, J. Holland, PVD CrxN coatings for tribological application on piston rings, *Surface and Coatings Technology*, 97, 1997, 661–668. DOI: 10.1016/S0257-8972(97)00335-6.
5. E. Broszeit, C. Friedrich, G. Berg, Deposition, properties and applications of PVD Cr N coatings, *Surface and Coatings Technology*, 115, 1999, 9–16. DOI: 10.1016/S0257-8972(99)00021-3.
6. A Barata, L Cunha, C Moura, Characterisation of chromium nitride films produced by PVD techniques, *Thin Solid Films*, 398–399, November 2001, 501–506.
7. S.M. Smith, S.A. Voight, H. Tompkins, A. Hooper, A.A. Talin, J. Vella, Nitrogen-doped plasma enhanced chemical vapor deposited (PECVD) amorphous carbon: processes and properties, *Thin Solid Films*, 398–399, 2001, 163–169.
8. K. Bouzid, N.E. Beliardouh, C. Nouveau, Wear and Corrosion Resistance of CrN-based Coatings Deposited by R.F Magnetron Sputtering, Tribology in Industry, 2015. http://hdl.handle.net/10985/10089.
9. W. Kern, Chapter II-4, Planar magnetron sputtering, In J. Vossen (Ed.), *Thin Film Processes*, Academic Press, Orlando, FL, 1979, p. 135.
10. J.L. Vossen, J.J. Cuomo, Chapter II-1, Glow discharge sputter deposition, In J. Vossen (Ed.), *Thin Film Processes*, Academic Press, Orlando, FL, 1978.
11. R. Wäsche, R. Ehrke, R. Kramer, A.R. Jayachandran, G. Brandt, Influence of temperature on tribological behaviour of DLC coatings under lubricated conditions up to 250°C, *Proceedings of Asia International Conference on Tribology 2018*, Malaysia, pp. 253–254, September 2018.

12. Donnet, C., & Erdemir, A. (Eds.), *Tribology of Diamond-Like Carbon Films: Fundamentals and Applications*, Springer Science & Business Media, Boston, MA, 2007.

13. J. Robertson, Diamond-like amorphous carbon, *Materials Science and Engineering: R: Reports*, 37(4–6), 2002, 129–281.

14. R. Waesche, M. Hartelt, V. Weihnacht, Influence of counterbody material on wear of ta-C coatings under fretting conditions at elevated temperatures, *Wear*, 267(12), 2009, 2208–2215.

15. D.X. Peng, C.H. Chen, Y. Kang, Y.P. Chang, S.Y. Chang, Size effects of SiO_2 nanoparticles as oil additives on tribology of lubricant, *Industrial Lubrication and Tribology*, 62(2), 2010, 111–120.

16. T. Singh, N. Kumar, J.S. Grewal, A. Patnaik, G. Fekete, Natural fiber reinforced non-asbestos brake friction composites: Influence of ramie fiber on physico-mechanical and tribological properties, *Materials Research Express*, 6(11), 2019, 115701.

17. M. Saleem, J. Ashok Raj, G. Shiva Sam Kumar, R. Akhila, Design and analysis of aluminium matrix composite spur gear, *Advances in Materials and Processing Technologies*, 0(0), 2020, 1–9.

18. P. Louda, S. Tumova, Z. Rozek, Application of nanotechnology at automotive industry, In *International Conference Vacuum and Plasma Surface Engineering*, Liberec, Technical University of Liberec, 2006, ISBN 80-7372-129-5.

19. J. Vetter, G. Barbezat, J. Crummenauer, J. Avissar, Surface treatment selections for automotive applications, *Surface & Coatings Technology*, 200, 2005, 1962–1968.

20. P. Forsberg, P. Hollman, S. Jacobson, Wear study of coated heavy duty exhaust valve systems in an experimental test rig. SAE Technical Paper, 2012-01-0546, 2012.

21. E.J. Mittemeijer, Fundamentals of nitriding and nitrocarburizing, In J.L. Dossett, E. Totten (Eds.), *ASM Handbook* Volume 4A, Steel heat treating fundamentals and Processes, ASM International, Materials Park, OH, 2013, pp. 619–646.

22. H. Holleck, Material selection for hard coatings, *Journal of Vacuum Science & Technology A*, 4(6), 2661, 1986.

23. J. A. Thornton, Influence of apparatus geometry and deposition conditions on the structure and topography of thick sputtered coatings, *Journal of Vacuum Science & Technology*, 11 (4), 666, 1974.

10 Tribology Aspects in Manufacturing Processes

Jyoti Menghani and Aayushi Meena
SVNIT

CONTENTS

10.1 INTRODUCTION

Manufacturing is defined in terms of transforming raw materials into a product. Among a wide variety of manufacturing processes available, the selection is based on criteria such as surface finish of the product, dimensional accuracy and geometrical

DOI: 10.1201/9781003097082-10

accuracy. The useful shapes of metal can be achieved by two manufacturing techniques, namely, primary manufacturing processes and secondary manufacturing processes. The primary manufacturing process includes casting and forming processes.

In the casting process, the molten metal is poured in die of the desired dimension, and the final product is obtained. However, it involves complex metallurgy of using molten metal. Casting processes can be classified into two broad categories:

- Expendable mould processes;
- Permanent mould processes.

The casting process selection depends on various factors including size, weight and complexity of the geometry, labour, equipment and tooling costs, tolerances and surface finish required, strength, quantity and production rate required and the overall quality requirements [1,2].

Metal forming is one of the oldest technologies known to humankind. In addition to zero wastage of raw material, other major advantages of forming over casting include rapid production rates and better mechanical properties. Nearly 90% of components in the car are made by metal forming processes. This is mainly due to advantages such as saving of material and labour cost, dimensional accuracy, prospect of generating the preferred internal stresses and of providing the objects with exceptional properties unreachable by other manufacturing methods, e.g., casting or machining

The process by which bulk metal is converted into a component or a part the known as metal working process. The process is classified into two modes: (1) process in which metal debris are produced (material removal operation like machining or metal removal processes) and (2) metal debris-free operations (metal-forming operation) [3].

The metal debris-free operation can be further classified as bulk metal forming (workpiece has high volume to surface area ratio), sheet metal forming (workpiece sheet has low volume-to-surface area ratio, usually thickness less than 6 mm) and sheet-bulk metal forming (wherein bulk-forming operations are applied on sheet metals) [4,5].

10.1.1 METAL WORKING OPERATIONS

Depending on the temperature of deformation, the forming process can be classified as either (1) hot working process or (2) cold working process. In hot working process, the temperature of deformation is higher than the recrystallization temperature (formation of strain-free crystals from strained crystal) of metals. Owing to high temperature, strain hardening does not occur, and large deformation is possible at lower forces as ductility is high and hardness is low. However, in the case of cold working since it is done at room temperature, metal gets hardened due to strain hardening. However, hot working involves high-energy consumption to heat the metal and dimensional tolerances and surface finish are poor. Cold working operations occur at room temperature. Compared to hot working, the dimensional tolerances are better and do not have metal loss due to oxidation [6].

Various Bulk Metal Forming Processes includes

a. Forging

Forging is the oldest method to form metals. It is a direct compression-type method in which the flow of metal is perpendicular to the direction of force applied. Because of orientation and refinement of grain structure, forged parts have higher strength than equivalent cast or machined part. Forging can be applied in both hot and cold conditions. The weight of forged part can vary from less than a kilogram to hundreds of metric tons. Examples of parts made by this process are as follows: wheel hub unit bearings, transmission gears, tapered roller bearing races, stainless steel coupling flanges, and neck rings for LP gas cylinders [7]. The schematic diagram of forging is shown in Figure 10.1a.

Compared to hot forging, cold forging has advantages of no scale formation, excellent tolerance and surface finish, thus eliminating any post forging secondary operations. Warm forging is carried out at an intermediate temperature between cold and hot forging, i.e., at temperature greater than room temperature but lower than recrystallization temperature. The dimensional tolerances are less as well as warm forging has poor surface finish compared to cold forging. However, it has reduced press load. In comparison with hot working, it has improved surface finish, thus reducing or eliminating finishing steps and close dimensional accuracy [8,9].

b. Rolling

Similar to forging, rolling is the process that uses direct compression to deform the metal to achieve desired size and shape. In rolling, the metal is passed through one or more pairs of rolls to reduce thickness (Figure 10.1b). The major benefit of rolling over other metal working process is that it is continuous working processes that do not require any dedicated dies and time-consuming setup operations and is characterized by high productivity and low cost.

To obtain the desired reduction in dimension in single pass, the largest quantity is first rolled by hot working followed by cold working. The initial material used in hot rolling includes 'raw material' such as large metal from unfinished casting; billets, slabs and booms that are processed and heated to a high temperature to convert to bars, rods, structural shapes, plates and sheets by incremental and successive cold rolling. After cold rolling, the products develop greater strength and wear resistance due to strain hardening. Thus, the actual process employed in the industry for the production of small gauge material is hot rolling to slightly above finished size required, cleaning/removing the oxidized surface by machining pickling or some other suitable process, and finally using cold-rolling of work-material to finished sizes. The applications of rolling include truck frames, automotive clutch plates, wheels and wheel rims, pipes and tubes. The literature indicates that theories of cold rolling depend on two physical characteristics, i.e., yield stress curve of strip and coefficient of friction between roll and strip [10,11].

The advancements in rolling include "rolling-extrusion" process whereby the combination of grooved rolls for rolling and equi channel step die is used for extrusion. This process results in making high-strength metal by reduction of grains to nanosized level scale, and this is of priority importance in the field of nanoscience. Another development is of slit rolling technology for the production of ribbed bars wherein the process uses special passes and guides to prepare, shape and separate the incoming billet into two or more individual strands. The process has advantages like easy adaptation to existing rolling mills with low investment costs, increase in production rates and reduction in a number of passes for rolling stands and operation costs. In the continuous roll-forming process, a non-uniform roll gap is formed by a pair of flexible rolls, and they rotate in opposite directions with identical angular velocity, the sheet metal is bitten into the roll gap by frictional force. The result is that the sheet metal is non-uniformly thinned in thickness direction and elongated in longitudinal direction. Thus, a 3D surface part with a desired shape is manufactured consecutively. In terms of material development, the extensive replacement of steel rolls with Tungsten carbide rolls in hot rolling industries (where steel wires are drawn in the intermediate and finishing stands) is adopted because of their excellent wear resistance and longer life of almost 50 times than steel rolls.

Another modification in rolling includes variable gauge rolling (VGR), wherein the thickness of the new products can be varied by adjusting the roll gap during the rolling process. In contrast to conventional rolling, there are three stages of VGR downward rolling, flat rolling and upward rolling. Flat rolling involves rolling in constant thickness portion and thickness transition regions with a variable thickness are formed by downward and upward rolling [12–16].

c. **Extrusion**

Extrusion occurs in a combined stress state similar to wire drawing, tube drawing, and deep drawing. It involves an indirect compression-type process wherein primary applied forces are tensile but due to the reaction of workpiece with die indirect compressive forces are developed. There are four classes of extrusion: direct extrusion; indirect extrusion; impact extrusion and hydrostatic extrusion

In direct extrusion, a solid ram drives the entire billet to and through a stationary die and must provide additional power to overcome the frictional power to overcome the frictional resistance between the surface of the moving billet and confining chamber (Figure 10.1c).

The major variables affecting extrusion pressure are die angle, reduction in cross-section, extrusion speed, billet temperature and lubrication. The two major advantages of extrusion process over other manufacturing processes are very complex; cross sections can be created, and since compressive and shear forces are involved, brittle material can also be processed. Other advantages include excellent surface finish and dimensional tolerances, amount of reduction in extrusion is very large and the process can be automated [17,18].

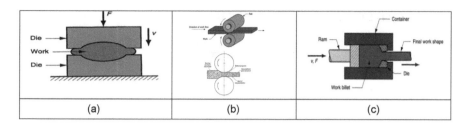

FIGURE 10.1 Principle of metal forming of metals [19]. (a) forging, (b) rolling, and (c) extrusion.

10.2 ROLE OF TRIBOLOGY IN METAL WORKING

For all the products of bulk metal forming, whether forging, rolling or extrusion, the dimensions and surface finish of the products are of utmost importance. All forming processes consist of three components (irrespective of forming conditions): work material, tool and (possibly) lubricant. Very aggressive conditions exist between workpiece and tool, resulting in high temperatures and pressures (may be local or uniform) at the interface. The surface of tool or die will worn resulting in deterioration of surface finish and dimensional tolerance. To use the tool for long term without being worn or scratch, there are high demands on the tool material and/or desired coating and/or need of lubricant in the working contact. Furthermore, friction plays an important role in energy consumption so by lowering friction; the environmental impact might be reduced. Therefore, research and development in regard to wear of the forming tools and consequently, tribology, is of great interest.

Tribology is an interdisciplinary field that involves not only the skills of mechanical engineers, materials scientists, physicists and chemists but also has expanded to medical and biological sciences including biotribology.

Tribology is the science and technology of interacting surfaces in relative motion (or of surfaces having a propensity to relative motion). Literature on the topic includes friction, wear and lubrication.

Tribology involves:

- Friction – the resistance to motion of one body moving against another. This resistance is a function of the materials, geometries and surface features of the bodies in contact, as well as the operating conditions and environment.
- Wear – the loss of material due to motion Wear is the loss of material, usually due to sliding. Typically wear is undesirable as it can lead to increased friction and ultimately to component failure.
- Lubrication – the use of a fluid to minimize friction and wear. Lubricants are primarily used to separate two sliding surfaces to minimize friction and wear. They also perform other functions, such as carrying heat and contaminants away from the interface [20].

  These three processes affect each other through interacting causes and effects, making the issue complex.

10.3 ROLE OF TRIBOLOGY IN INDUSTRY

i. To conserve material resources through reduction of wear,
ii. To conserve energy through reduction of friction,
iii. To increase the operating life of the equipment through maintenance engineering,
iv. To increase standards on product reliability and personnel safety.

10.3.1 FUNDAMENTALS OF TRIBOLOGY

There are various manmade and natural systems in which there is a relative motion of interacting bodies. Friction, which is due to wear and energy dissipation, is responsible for consuming about one-third of the world's energy resource and attempt to overcome its barrier in one form or another [21].

a. Friction:
Friction is the force resisting the relative motion of solid surfaces, fluid layers, and material elements sliding against each other. In addition, in the manufacturing process, friction results in complex physicochemical processes proceeding in the contact zone of rubbing bodies.

Broadly, friction can be classified into two main types:

i. **Dry Friction**: as the name indicates it exists where there is an interaction of rubbing bodies in absence of any fluids/lubrications, which involves (a) static and (b) rolling friction

ii. **Wet Friction**: There is the presence of fluid between interacting surface and can be further classified into (1) fluid friction exists in the presence of lubricant between rubbing bodies, (2) mixed friction it occurs during the simultaneous presence of dry and fluid friction regions in the contact zone; and (3) boundary friction when the lubricant thickness is smaller than ten atomic layers or two rubbing surfaces separated by a lubricant layer interact with each other due to asperities or roughness boundary friction occurs [22,23].

Thus, friction is not a material property; it is a system response in the form of a reaction force. The relative ease with which materials slide over one another is defined in terms of coefficient of friction. It is a dimensionless quantity and is defined as the ratio of two forces acting, respectively, perpendicular and parallel to an interface between two bodies under relative motion or impending relative motion. There are two types of friction coefficient

1. Static friction coefficient that opposes the onset of relative motion (impeding motion) and

2. Kinetic friction coefficient that opposes continuation of relative motion once the motion has started. Drum and disk brakes or clutches or drives are examples in which performance is dependent not only on the transition from static to kinetic friction but also on static and kinetic friction individually.

$$\mu_s = F_s / P \tag{10.1}$$

$$\mu_{sk} = F_k / P \tag{10.2}$$

The force F_s is sufficient to prevent the relative motion between two bodies and F_k is force needed to maintain relative motion between two bodies and P is force normal to interface. The static friction coefficient is at the time of stationary contact, whereas kinetic friction coefficient depends on sliding velocities. The static friction is time-dependent [24–26].

Friction between a tool and the workpiece has a significant effect on the material deformation, forming load, component surface finish and die wear. The coefficient of friction, if controlled properly, could generate the required stresses to deform the metal to the desired shape. It could also lead to fracture of the sheet if not controlled properly.

As reported in [27], in the case of forming surface texture of die plays an important role. Two sets of dies with variation in grinding marks were used, and this resulted in variation in forging loads. The variation in grinding marks resulted in variation in friction conditions at interface and hence metal flow pattern. As per reference [28], in upset forging, various machining processes such as grinding, milling, electrospark machining and lathe turning were applied o flat die surface to vary frictional condition and study its effect on barrelling. In addition to this, various researchers have performed numerical stimulation as well as using the FE tool to determine the role of friction on metal working processes such as deep drawing process, cold upsetting-extruding of tube flanges.

Amongst various factors that affect frictional behaviour during metal forming, material (composition, melting point and surface finish) contribute maximum to frictional force. Other important factors include surface roughness and asperity interlocking. The presence of high friction at the interfaces between surface results in a large amount of heat and energy dissipation [29]. One of the most effective ways to control friction is to use a lubricant in liquid and/or solid forms. Even certain gases may be used to lubricate rubbing surfaces. Liquid lubricants reduce friction by preventing sliding contact interfaces from severe or more frequent metal-to-metal contacts or by forming a low-shear, high boundary film on rubbing surfaces [30].

b. **Lubrication**:

Depending on the process selected whether it is cold working or hot working with the purpose to give the desired shape of component localized compressive forces using specific forming tools is done. In certain hot forming processes depending on metal composition, the metal working process may involve high temperature (@750°C) and high pressure (up to 1379 MPa) and is associated with high friction and tool wear. Owing to friction, high temperature is generated during processing. Under these circumstances, the problem of high temperature can be resolved by applying lubricant fluids. Lubricants that are applied in metalworking operations are known as metalworking fluids (MWFs). MWF reduces friction between tool and workpiece, thus reducing the amount of heat generated. Furthermore, the heat generated during the process is dissipated, resulting in further cooling and avoiding thermal damage of workpiece and tool wear. In addition to reduce wear and friction between rotating and stationary components, additional functions of lubricant include shock absorption, damping of noise, minimizing corrosion, flushing contaminant and acting as sealing media [31].

Since various mechanical components work under varying operating conditions due to change in lubrication condition, there is significant variation in interfacial friction. The Stribeck curve gives the general aspect of variation of friction over the transition of lubrication modes. The curve was first invented by Richard Stribeck (1902) confirmed the existence of minimum friction for journal bearing friction experiments. It was then modified by Mayo Hersey in which they demonstrated that friction due to viscous shear was a unique function of the product of viscosity (μ) by rotational speed (N) divided by the average load (P), which is called the Hersey number, $\mu N/P$. The friction coefficient plotted as a function of the Hersey number is known as the Stribeck curve or Lambda curve. Originally, it described frictional characteristics in journal bearings and now it is extended to other types of contact geometry found in engineering applications.

The Stribeck curve is widely used for the identification of friction and wear in different lubrication regions. Each lubrication region is characterized by definite values of friction, wear and structural state of surface layers. Thus, the overall view of friction variation in an entire range of lubrication is obtained by Stribeck curve. Figure 10.2 shows a schematic Stribeck curve for a journal bearing system, where regimes of lubrication, the full film, mixed and boundary film lubrications are also illustrated.

Boundary film lubrications
In this region, the coefficient of friction is high and constant and does not vary with operating conditions as speed and load. The friction is dominated

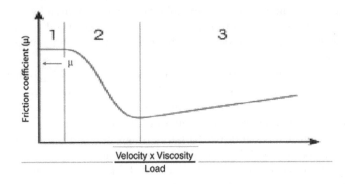

FIGURE 10.2 Schematic Stribeck Curve (Hersey number on horizontal axis, Friction coefficient on vertical) 1. Boundary Lubrication, 2. Mixed Lubrication and 3. Hydrodynamic Lubrication.

by surface and boundary film characteristics. Here load is carried by direct tool workpiece contact at asperities level.

Mixed Lubrication
In this case, the load is shared by lubricant film and asperity contacts. As seen in Figure 10.2 with an increase in speed, friction decreases because of the increase in thickness of lubricant.

Hydrodynamic Lubrication
In this case, the thickness of film is much greater than surface roughness. Initially, in this region, the friction reaches to a minimum and then slowly increases due to shear strain rate increase and friction originates from bulk rheological properties of lubricants at very high pressure and strain [32–36].

Lubrication is classified as partial when there is generally between 0.01 and 0.1. When there is a total separation of the block and the plate by a layer of lubricant, it is called fully hydrodynamic. The coefficient of friction for fully hydrodynamic lubrication is usually between 0.001 and 0.01 [37]

10.4 CLASSIFICATION OF METAL WORKING FLUIDS

There are various classification of lubricants however lubricants used for MWF is classified as per DIN 51385 shown in Figure 10.3. MWF are classified into two classes: water-based and oil-based [3].

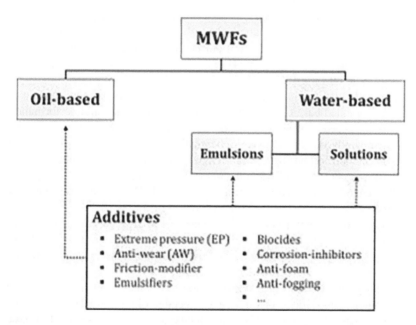

FIGURE 10.3 Classification of Metal Working Fluids types according to DIN 51385.

10.4.1 FORMULATION OF LUBRICANT (LIQUID (OIL-BASED OR WATER-BASED, SEMI-SOLID AND SOLID OR DRY)

Water-based MWF is used where heat dissipation is more important than lubrication, whereas in the reverse case is true for oil lubrication. Solid lubricants are solid powders and most commonly graphite or MoS_2 that have layered structure and weak interlayer van der Waals force which gives high compression strength in the direction perpendicular to sliding movement and low shear strength in the direction of the sliding movement giving property of lubricity. The advantages of solid lubricant include high-temperature stability, and chemical inertness does not require any sealing and no contamination. Hence, under extreme working conditions such as elevated temperatures, high pressures, cryogenic temperature or under vacuum and radiation, solid lubricants are widely used.

However, graphite and molybdenum disulphide tend to blacken the machines and operators clothing hence other compounds sulphides (zinc sulphide), metal hydroxides (calcium hydroxide), phosphorus compounds, (zinc pyrophosphate or calcium carbonate)oxides (calcium oxide and lead oxide), are considered as solid additives as well [38–40].

Further, oil-based lubricant can be classified as:

a. **Mineral oil-based lubricants**

Mineral oil is one of the most widely used base oils in the lubricant industry. It is a liquid by-product of refining crude oil to make gasoline and other petroleum products. Mineral oil has various advantages such as high chemical stability, good thermal and oxidation resistance and low price.

Mineral oil contains various toxic materials like heavy metal benzene series and polycyclic aromatic hydrocarbons (PAHs). The used mineral oil causes serious harm to the environment and human health if discharged directly into the environment [41,42].

b. **Synthetic oil-based lubricants**

Synthetic oils are compared to mineral oil that contains more amount of highly refined base oils, resulting in superior protection and performance. It is produced by chemical synthesis, and hence, the composition can be scientifically designed through chemical combination as per requirement The synthetic oil has better high-temperature stability and oxidation stability, lower coefficient of friction and better low-temperature viscosity [43].

c. **Biodegradable, environmentally friendly oils (based on esters or fatty oils)**

With increased environmental awareness as a large portion of lubricants end up in polluting the environment and the purpose to initiate and boost use of biodegradable products, various government incentives and mandatory regulations are in practice.

There are five biodegradable base stocks of some significance. These are (1) highly unsaturated or high oleic vegetable oils (HOVOs), (2) low viscosity polyalphaolefins (PAOs), (3) polyalkylene glycols (PAGs), (4) dibasic acid esters (DEs) and (5) polyol esters (PEs)

1. *Vegetable oils*

   Vegetable oils have a major drawback that they cannot be used at high temperature because they have tendency to oxidize at high temperature. However, they are less toxic to environment and human beings since they are derived from renewable resources and do not contain volatile organic chemicals [44,45].

2. *Polyalphaolefins (PAOs)*

   The biodegradable PAOs are the low-molecular dimers to tetramers. Because of their low molecular weight, they can be used for low-temperature applications. Hence, they are widely used as hydraulic and engine oil, particularly in cold climate applications and where hydraulic pressures are increasing (>7000 psi).

3. *Polyalkylene glycols (PAGs)*

   Polyalkylene glycols are produced by polymerizing ethylene oxide and propylene oxides or their mixtures. They are highly polar; hence, they have high solubility of water and which is advantageous to a biodegradable fluid but disadvantageous as it intensifies moisture contamination.

4. *Diesters (DEs)*

   The diesters are composed of dicarboxylic acids (such as adipic acid) esterified with alcohols produced from hydroxylated petroleum fractions. It has the advantage of use as base stocks for biodegradable hydraulic fluids and can be mixed with PAOs for excellent properties.

5. *Polyol esters (PEs)*

   Polyol esters are composed of fatty acids attached to alcohol, which does not contain a hydrogen atom on the β-carbon atom. Owing to typical structure, they have high resistance to oxidation, excellent low-temperature performance and high biodegradability.

10.4.2 MANUFACTURING PROCESS ADOPTED (CUTTING FLUID, GRINDING FLUID OR FORMING FLUID)

Lubricants are used in metal forming processes to provide desirable tribological conditions at tool–work piece interface, increase metal formability, extending tool life and reducing energy loss.

The lubrication affects material formability and forming defects throughout forming procedure, hence the feasibility and productivity of a forming process depend strongly on the provided lubrication. Load required for the deformation during forming, life of process tool and quality of part produced are other critical factors that can be influenced by lubricant performance [46,47]

Various process parameters as load, temperature and speed of deformation in addition to inherent lubricating properties of lubricant play an important role in the selection for the particular metal working process.

As discussed based on metallurgical considerations, the metal working processes can be classified into three processes: hot working, i.e., processing above the recrystallization temperature, cold working, i.e., processing below recrystallization temperature and warm working processing at 50% of recrystallization temperature.

Since this classification is based on melting point of metal, it is not suitable from tribological considerations. For tribological purposes, it may be better to consider cold-forming processes as those operating below about 250°C, where organic-based lubricants can be used without significant thermal degradation

Since in bulk-forming processes, as rolling, extrusion, and forging the interface pressure (between tool and workpiece) equivalent to tensile strength same as the material to be processsed is applied. The details are discussed in tribology at the high-temperature section [48].

10.4.3 QUANTITY OF FLUID

Several important processes like cold forging and hot extrusion or hot rolling of steel are possible only by using an efficient lubrication system. In the case of cold forging phosphate stearate lubricant coatings, hot extrusion molten glass film lubricants and for hot rolling of steel, sodium bicarbonate and sodium nitrate are used. In the case of cold rolling of Al, the production speed and surface quality are dependent on control of lubrication.

The quantity of lubricant used is dependent on viscosity of lubricant,t which in turn is dependent on temperature, pressure and rate at which it is sheared. A thinner layer will be created with less viscous lubricant, which will result in more metal-to-metal contact and hence higher friction and wear. On the other hand, more viscous lubricants create thicker layer, less metal-metal contact, low friction and lesser wear.

Moreover, the performance of lubrication is influenced by other factors like type of manufacturing processes, material being processed and tool material used.

The lubrication in metal working involves different mechanisms, hence its selection depends on the chemistry of the metal working tool-lubricant-work piece interface, the method of lubricant application, the geometry of the process, ease of disposal, the mechanics of the operation [3,49].

10.4.4 ADDITIVES FOR LUBRICANTS

The purpose of additives is to modify the properties of lubricants and make them suitable for particular applications. The additives and their characteristics and functions are mentioned in the table below. Additive selection depends on the requirements for a particular tribological system [50].

In addition, the adequate choice of additive depends on the specific lubrication regime (see Figure 10.1): for boundary lubrication conditions (mostly at high loads), extreme pressure additives (EP) are more effective; anti-wear (AW) and friction modifiers (FM) are more efficient in the mixed lubrication regime; and finally, viscosity improver additives (VII) are more effective at the full film lubrication regime [51–53].

10.5   WEAR

During forming the tool surface is subjected to a sliding contact which results in loss of material from the surface and transferred to another surface. The high contact stresses and often elevated temperature leads to tool wear. The wear mechanisms that

TABLE 10.1

Role of Additives in Lubricants

Additives	Key Characteristics
Anti-corrosion	Reduces deterioration of active metals
Anti-oxidant	Increase oxidation resistance and thus prolongs life of base oil
Anti-rust	Slows corrosion of iron alloy
Anti-wear	Forms protective film to protect the metal surface
Colour/ultraviolet dye	Visual markers for inspection or assembly
Conductive agent	Add thermal or electrical conductivity
Extreme pressure	Forms protective layer to prevent seizure and severe damage
Friction modifier	Reduce the coefficient of friction
High-temperature enhancer	Boosts high-temperature limit of oil
Viscosity improver additives (VII)	Alters oil viscosity

can be found in forming applications include abrasive and adhesive wear, mechanical and thermal fatigue, plastic deformation and corrosion. For excellent wear resistance, the surface of the tool should be hard and capable to maintain high hardness at high temperature, especially for high abrasion wear resistance. Another important characteristic of tool material is that it should have high toughness so as to improve fatigue properties and high thermal resistance.

10.5.1 WEAR DURING COLD FORMING

Since cold forging is forging at room temperature the tool failure is due to high contact pressure and relative motion between tool surface and formed material. Among various wear mechanisms like abrasive wear, adhesive wear, low-cycle fatigue, crack propagation and plastic deformation most dominant is adhesive wear. Very often, more than one wear mechanism can occur simultaneously.

There are local micro welds between tool surface and workpiece and later on accumulation of work material on tool surface is the main reason for adhesive wear.

The outcome of this is poor uneven surface due to unstable friction and scratching. Removal of adhered work material may cause three-body abrasive wear. The most common method to avoid adhesive wear is to use hard and tough tool material with excellent surface finish. Application of lubricants reduces/avoids adhesive wear. During forming of hard material or multiphase material(consisting of oxides, carbides, etc), abrasive wear occurs. The reason is hard material causes scratch on tool surface and leads to tool wear. Abrasive wear can be found in punching, fine blanking, drawing, extrusion and forging. The abrasive wear can be avoided by using the tool which is surface engineered either by or in combination of thermo-chemical treatments and /or the deposition of hard, wear-resistant coatings.

10.5.2 WEAR DURING HOT FORMING

In contrast to cold forming hot forming involves high processing temperature. The wear of tool occurs by four types of loading viz thermal, mechanical, tribological and chemical. Due to high temperature and Cyclic heating and cooling during processing (thermal fatigue) tool hardness is reduced. In addition, due to high temperature, there is hard and brittle oxide formation on the surface of metal which in turn generates hard abrasive particles and causes abrasive wear of tool. There will be a larger drop in surface hardness and tool will be more reactive at high temperature, which would intensify abrasive wear and galling.

The understanding of tool wear in forming will assist in the selection of tool material and /or surface treatment required as per reference [54]. The costs of roll wear are estimated to be 10% of processing costs [53,55]

10.6 TRIBOLOGY AT HIGH TEMPERATURE

When the metal surface is exposed to high temperature, various changes are occurring on the metal surface due to oxidation and diffusion which results in changes in morphology, microstructure and mechanical properties. These changes in turn will affect the tribological behaviour of materials at elevated temperatures. Another important aspect is poor performance of lubricant at temperature greater than 300°C [56]. Tribology at high temperature forming mainly concentrates on tool–workpiece interface. Four main topics in hot forming are considered:

1. Oxidation at high temperature and its effect on tribology behaviour
2. Surface coating and their tribological behaviour
3. Lubrication effect and its tribological performance at high temperature
4. Testing method that is tribometers in metal forming at elevated temperature.

10.6.1 OXIDATION AT HIGH TEMPERATURE AND ITS EFFECT ON TRIBOLOGY BEHAVIOUR

The formation of oxides on metal surfaces in relative motion often results in reduced wear and friction. Hence, such oxides can play an important role in many practical situations. Considering the case of Ni and Co which are widely used metals for hot metal forming because of their high melting point and retain high strength at high temperature. In this case, very fine crystalline particles of oxides such as NiO, Cr_2O_3, Co_3O, $NiCr_2O_4$, etc are formed. These particles tend to deform plastically during sliding and produce smooth surface. The iron-base alloys especially low alloys steels are widely used because of lower cost. However, when subjected to high temperature (greater than 200°C) environment they suffer from oxidation. During hot working processes (forging, rolling), oxidation occurs on the surface which deteriorate the properties of metal and is to be removed during subsequent cold working or further applications. There is variation in the composition of iron oxide over a range of temperatures. From 200°C to 570°C two-layer oxide scale, outer scale is Fe_2O_3 (heamatite) and inner layer is of Fe_3O_4 (magnetite). This duplex layer gives good protection against further oxidation as it inhibits diffusion of iron and oxygen ions. Above 570°C third iron oxide, FeO (wustite) is formed at the metal/Fe_3O_4 interface and the oxidation rate increases due to the higher defect concentration of this layer. Hence, for high-temperature applications, the alloy steel consisting of elements that form stable/slow growing barrier layer at the base of oxide scale, e.g., Cr_2O_3 layer in a Ni-20wt%Cr alloy [56–59].

In addition to temperature composition of steel also plays a major role in determining the rate of oxidation and stability of the oxide layer. In the case of plain carbon steel at a temperature below 500°C, increasing carbon content increases the rate of oxidation, however, at a temperature greater than 700°C rate of oxidation is reduced. When alloying elements as Cu and Ni are added (more noble than iron), a strong bond is formed which makes the scale difficult to remove between reheating and hot forming [60].

10.6.2   SURFACE COATING AND THEIR TRIBOLOGICAL BEHAVIOUR

Maximum damage occurs at tool–workpiece interface during hot metal forming process due to any combination of reasons like galling (material transfer), mechanical loads, thermal fatigue, oxidation, plastic deformation and cracking. Since all these are surface phenomena, if the protective coating is applied on the surface of the tool and workpiece, it will provide resistance to wear, fatigue and oxidation. In addition to coating application, optimized heat treatment of tool steel may also improve the properties of tool steel [60].

As per reference [61], they selected martensitic chromium hot work tool steel to prevent damage due to crack formation during thermal fatigue. These steels have characteristics of high hot strength, toughness and ductility and adequate thermal conductivity and expansion. Heat treatment determines thermal fatigue crack and test temperature determines thermal fatigue resistance.

The investigations of various types of surface modifications like different grades of polishing, plasma nitriding and DLC coating, all applied to cold work tool steel for their tribological behaviour were done [62]. The important observation made was that surface roughness plays a major role since it removes all surface irregularities and hence potential source of metal transfer. DLC carbon-based coating maintains low friction and galling resistance. However, the performance of fine polishing is better than DLC. In the case of plasma nitriding, additional performance of TiN coating was poor, and it showed high friction and high affinity to pick up work material [58,61,62].

In reference [63] to compare thermal diffusivity and wear performance, five PVD films of different compositions, TiN, TiAlN, CrN, AlTiN, and TiCN, were deposited on M41 steel. The results suggested that among all the coatings TiAlN and AlTiN performed better than the remaining coatings. The work done by Hardell and Prakash was to study high-temperature tribological applications of potential PVD coating (TiAlN and CrN) and plasma nitriding treatment against ball bearing steel at 400°C and found out that TiAlN performed better than CrN.

10.6.3   LUBRICATION AT ELEVATED TEMPERATURE

Owing to complex tribological conditions like oxidation on the metal surface of metal or tool or both, wear or thermal load the role of lubricant is to protect freshly exposed chemically active metal surface. Certain research in high-temperature lubrication is discussed [64].

As per reference [65] the lubricant used for hot forging should have a high ignition point, a proper viscosity at the working temperature and a low coefficient of friction, protect the tool from overheating and prevent a temperature drop of the forging (it should characterize in low thermal conduction) and reduce the coefficient of friction during the release of the forged metal from the die cavity. The authors have given comprehension studies for the same.

The paper [65] presents the influence of the applied lubricating and cooling agents, lubricant dose size, time and direction of administration as well as methods of operating the lubricating devices for high-temperature forging operations. As it has been demonstrated, the selection of the lubricant, its dose and method of feeding continues to be a current research problem. They concluded that lubricants most commonly used in die forging at elevated temperature are graphite-based lubricants with modified properties depending on the specific process.

In reference [66], Al-coated 22MnB5 steel was selected and a comparison of coefficient of friction was done using five types of commercial lubricant for hot forging using a hot flat drawing stimulator. Different newly developed solid lubricants were studied. In this study, first, the coefficients of friction of Al-coated 22MnB5 steel were measured using five types of commercial lubricant for hot forging using a hot flat drawing tribosimulator. Second, from the results, the coefficient of friction was measured using five types of lubricant with added different solid lubricants at a concentration of 5% (solid lubricants, namely, swellable and nonswellable mica, melamine cyanuric acid, potassium titanate, and cellulose powder). Finally, the efficiency of newly developed lubricants was determined using the hot deep drawing test machine. The results indicated that the addition of hydrophilic polymer and a mineral salt resulted in the lowest coefficient of friction for hot forging whereas swellable micagace lowest value for hot stamping.

In the research of reference [67] lubrication behavior of graphite was investigated by ring compression tests of $Ti\pm6Al\pm4V$ alloy at temperatures from 750°C to 1000°C and constant strain rates from 0.05 to15s^{-1}. The results indicated that Graphite powder dispersed in engine oil can be used effectively as a lubricant for compression tests at temperatures below 800°C above this temperature the lubricating properties are not very effective.

In the research [68], different lubricants (mineral oil, alcohol, ester, phosphor EP, sulphur EP, and MoS_2) were tested in a range of 200°C–300°C in a ring compression test of Mg alloy. Among all the lubricants, thin alcohol was the most efficient in reducing friction under the prevailing testing conditions

10.6.4 TESTING METHOD (THAT IS TRIBOMETERS IN METAL FORMING AT ELEVATED TEMPERATURE)

Friction is one of the most important parameters in tribology that determines the behaviour of the material during metal working. There are two methods of determining friction: direct and indirect. In direct method, measurement pins are placed locally at tool–workpiece interface. Indirect testing method involves friction measurement due to friction stresses, which are averaged over an entire tool–workpiece interface. The processes have advantages of simple process, represents actual conditions compensates local inhomogenieties hence widely used in industry. However, these processes have disadvantages that most of the parameters cannot be controlled independently from each other [69].

With the purpose of energy saving, Al extruded parts are widely used. Since hot working increases the production rate, hence it is a widely preferred method. As per reference [70], the friction at the workpiece–tooling interfaces during hot extrusion of aluminium (temperature range from 300°C to 500°C) is one of the important boundary conditions, and it is very difficult to determine. During extrusion, complex tribological phenomena occur and are affected by local temperature, relative velocity, geometry and tooling surface roughness and all this is more important because the tool is made of steel and has to come in contact with hot aluminium. Under the effect of friction, severe plastic deformation, removal of heat from extrusion tooling to surrounding there is variation in temperature during extrusion; in addition, there is a normal pressure variation due to adhesion friction and varied velocities. All these phenomena affect the surface topography in a complex way at workpiece tooling interface. Usually, to determine the tribological behaviour of a mating system and understand the mechanisms of wear and friction, the most commonly used characterizing test are pin-on-disc and ball-on-disc friction tests and extending these tests at high temperature resembles the contact conditions between the billet material (an aluminium alloy) and extrusion tooling (made of a steel).

Another most common method of metal forming, i.e., hot rolling is considered. There are three main components of hot strip mill viz the Mill, the rolled metal and interface. The tribology at interface involves friction, lubrication and heat transfer. Various material parameters include mechanical properties of rolled material, rolls including yield stress, hardness, Young's modulus, shear modulus as well as thermophysical properties affect the interactions. Since the process occurs at high temperature, various surface properties such as the chemical reactivity and nature of scale formed (which in turn depends on chemical composition) at high temperature as well as surface roughness plays role in tribological performance. Additionally, the process parameters like temperature, speed, and extent of reduction will influence surface interactions [54,71].

10.7 TEST TO STIMULATE HOT WORKING PROCESSES

10.7.1 RING COMPRESSION TEST

Planestrain or uniaxial conditions exist during the ring compression test, hence it is most suitable for measuring the material's flow strength during bulk forming (compressive process). However, ring compression test has inherent disadvantage of inhomogeneous deformation within the sample, caused by the undesirable specimen/die interface friction. In ring compression test, hollow cylinder is axially deformed between two flat dies. The principle of the test is different behaviour of inner diameter. At low friction, inner and outer diameter deform symmetrically and material flows in an outward direction. At high friction, opposite behaviour is observed and inward material flow occurs. The relative change in inner diameter and ratio of compression after test is plotted graphically and friction parameter either friction coefficient μ or

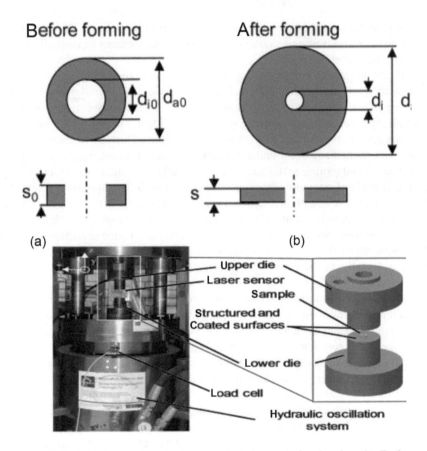

FIGURE 10.4 (a). Ring compression test sample before and after forming. (b). Tool system for the ring compression test with the hydraulic oscillation system.

the friction factor m is determined. The schematic diagram of ring in compression test is shown in Figure 10.4a and b [72–74].

As shown in Figure 10.5, the height reduction during compression test in terms of percentage and corresponding measurement of internal diameter of test specimen provides quantitative knowledge of the magnitude of the prevailing friction coefficient at the die/workpiece interface [75].

10.7.2 Block on Cylinder Test

The block on cylinder test represents laboratory-scale extrusion metal working process. This process is adopted as field test, which is time-consuming, expensive and difficult to control. The schematic diagram is shown in Figure 10.6. As per simulation, extrudate is rotating cylinder and contact surface of block represents bearing of the tool. As per ref. [75], the laboratory test represents similarities to actual extrusion in terms of wear observed in both cases that is combination of abrasive, chemical and delamination wear. In their investigation, they had done a comparison between

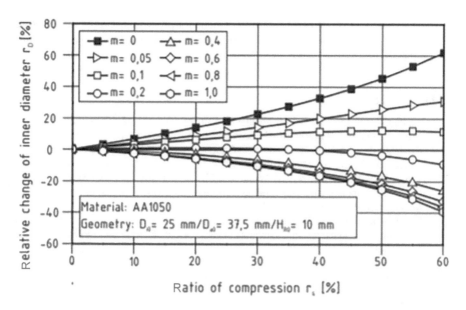

FIGURE 10.5   Graph of the ratio of compression to a relative change in diameter in ring compression test.

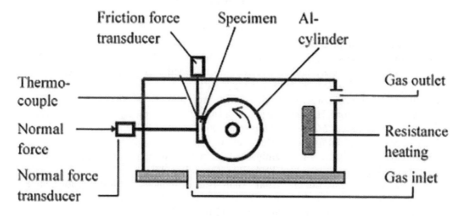

FIGURE 10.6   Schematic of test equipment.

traditional nitriding and a CVD coating by TiC+TiN on die. They concluded that chemical wear dominates the deterioration for the two surface treatments for both cases. However, the CVD coating had a superior resistance to wear compared to the nitrided steel.

10.8   SUMMARY

The chapter deals with most common bulk metal forming processes including forging, rolling and extrusion and details about tribological behaviour in terms of friction, wear and lubrication. Details about various types of lubricants and additives are also discussed. The variation in wear phenomena for processes at room temperature and high temperature are analyzed. Finally, two laboratory scale tribometers replicating the actual conditions are also discussed.

REFERENCES

1. Scallan, P. Chapter 4-Material evaluation and process selection. Peter Scallan (Ed.), *Process Planning*, Butterworth-Heinemann, Burlington, MA, 2003, pp. 109–170.
2. Gronostajski, Z., Pater, Z., Madej, L., Gontarz, A., Lisiecki, L., Lukaszek-Solek, A., Luksza, J., Mróz, S., Muskalski, Z., Muzykiewicz, W., Pietrzyk, M., Sliwa, R.E., Tomczak, J., Wiewiórowska, S., Winiarski, G., Zasadzinski, J., Ziólkiewicz, S. Review recent development trends in metal forming. *Archives of Civil and Mechanical Engineering* 19 (2019) 898–941.
3. Osama, M., Singh, A., Walvekar, R., Khalid, M., Gupta, T.C.S.M., Yin, W.W. Recent developments and performance review of metal working fluids. *Tribology International* 114 (October 2017) 389–401.
4. Dixit, P.M., Dixit, U.S. *Modeling of Metal Forming and Machining Processes*, Springer-Verlag, London, 2008, p. 2. Hardcover ISBN 978-1-84800-188-6, Softcover ISBN 978-1-84996-749-5.
5. Merklein, M., Gröbel, D., Löffler, M., Schneider, T., Hildenbrand, P., (2015). Sheet-bulk metal forming – forming of functional components from sheet metals. *MATEC Web of Conferences*. 21. 01001. doi: 10.1051/matecconf/20152101001.
6. https://en.wikipedia.org/wiki/Hot_working#cite_note-Degarmo373-1.
7. https://en.wikipedia.org/wiki/Forging#:
8. Lee, Y., Yoon, E., Nho, T., Moon, Y. Microstructure control of ferrous driven part fabricated by warm precision forging. *Procedia Manufacturing* 15 (2018) 404–410.
9. Shivpuri, R., Babu, S., Kini, S., Pauskar, P., Deshpande, A. Recent advances in cold and warm forging process modeling techniques: Selected examples. *Journal of Materials Processing Technology* 46 (1994) 253–274.
10. https://en.wikipedia.org/wiki/Rolling_(metalworking).
11. Whitton, P.W., Ford, H. Surface friction and lubrication in cold strip rolling. *Proceedings of the Institution of Mechanical Engineers* 169(1) (1955) 123–140.
12. Kliber, J. Advanced forming technology. *Metalurgija* 55(4) (2016) 835–838.
13. Turczyn, S., Dziedzic, M., Kuźmiński, Z. A study on design of slitting passes used for rebar rolling, May 21st – 23rd 2014, Brno, Czech Republic, EU.
14. Wang, M., Liu, Zn., Lu, G.L. et al. Analysis of continuous roll forming for manufacturing 3D surface part with lateral bending deformation. *International Journal of Advanced Manufacturing Technology* 93 (2017) 2251–2261.
15. Shafiei, E., Dehghani, K. Effects of deformation conditions on the rolling force during variable gauge rolling. *Journal of Manufacturing and Materials Processing* 2 (2018) 48.
16. Kumapay, O.M.I., Akinlabi, E.T., Onu, P., Abolusoro, O.P. Rolling operation in metal forming: Process and principles – A brief study. *Material Today Proceeding* 26(Part 2) (2020) 1644–1649.
17. Yadav, R.R., Dewang, Y., Raghuvanshi, J. A study on metal extrusion process. *International Journal of LNCT* 2(6) (2018) 124–130.
18. https://en.wikipedia.org/wiki/Extrusion.
19. https://www.slideshare.net/aslam1992/u4-p1-metal-forming-processes.
20. Zhang, C. Understanding the wear and tribological properties of ceramic matrix composites. *Advances in Ceramic Matrix Composites* (2014) 401–428.
21. Amiri, M., Khonsari, M.M. On the thermodynamics of friction and wear—A review. *Entropy* 12 (2010) 1021–1049.

22. Lyashenko, A. Tribological properties of dry, fluid, and boundary friction. *Technical Physics* 56(5) (2011) 701–707.
23. https://www.slideshare.net/MikeRamsey17/tribology-wear-and-friction-an-overview
24. Persson, B.N.J., Albohr, O., Mancosu, F., Peveri, V., Samoilov, V.N., Sivebaek, I.M. On the nature of the static friction, kinetic friction and creep. *Wear* 254 (2003) 835–851.
25. Hwang, D.-H., Zum Gahr, K.-H. Transition from static to kinetic friction of unlubricated or oil lubricated steel/steel, steel/ceramic and ceramic/ceramic pairs. *Wear* 255(1–6) (August–September 2003) 365–375.
26. Blau, P.J. The significance and use of the friction coefficient. *Tribology International* 34(9) (2001) 585–591.
27. Menezes, P.L., Kumar, K.,Kailas, S.V. Influence of friction during forming processes— a study using a numerical simulation technique. *The International Journal of Advanced Manufacturing Technology* 40(11–12) (2008) 1067–1076.
28. Malayappan, S., Narayanasamy, R. An experimental analysis of upset forging of aluminium cylindrical billets considering the dissimilar frictional conditions at flat die surfaces. *International Journal of Advanced Manufacturing Technology* 23(9–10) (2004) 636–643.
29. Che, Q., Li, H., Zhang, L., Zhao, F., Li, G., Guo, Y., Zhang, J, Zhang, G. Role of carbon nanotubes on growth of a nanostructured double-deck tribofilm yielding excellent self-lubrication performance. *Carbon* 161 (2020) 445–455.
30. Berman, D., Erdemir, A., Sumant, A. V. Graphene: A new emerging lubricant. *Materials Today* 17(1) (2014) 31–42.
31. https://www.industrial-electronics.com/engineering-industrial/rotating-2.html.
32. Wang Y., Wang Q.J. Stribeck Curves. In: Wang Q.J., Chung YW. (eds) Encyclopedia of Tribology. Springer, Boston, MA. 2013, pp 395–400, 2013.
33. Wang, Y., Wang, Q.J., Lin, C., Shi, F. Development of a set of stribeck curves for rough surfaces. *Tribology Transactions* 49(4) (2006) 526–535.
34. Moshkovicha, A., Perfilyeva, V., Gorni, D., Lapsker, I., Rapoport, L. The effect of Cu grain size on transition from EHL to BL regime (Stribeck curve). *Wear* 271(9–10) (29 July 2011) 1726–1732.
35. Guegan, J., Kadiric, A., Gabelli, A., Spikes, H. The relationship between friction and film thickness in EHD point contacts in the presence of longitudinal roughness. *Tribology Letters* 64 (2016) 1–5.
36. He, T., Zhu, D., Wang, J., Jane Wang, Q. Experimental and numerical investigations of the stribeck curves for lubricated counterformal contacts. *Journal of Tribology* 139(2) (July 2016), 021505-1–021505-13.
37. McKeen, L.W., Chapter 2- Introduction to the tribology of plastic and elastomers, In McKeen, L.W., Editor, *Fatigue and tribological properties of plastics and elastomers*, 2nd ed., William Andrew, Amsterdam, Oxford, 2010. pp 28–38.
38. Kajdas, C. Additives for metalworking lubricants - a review. *Lubrication Science* 1(4) (1989) 385–409.
39. Kumar, R., Banga, H.K., Singh, H., Kundal, S. An outline on modern day applications of solid lubricants. *Materials Today: Proceedings* 28(Part 3) (2020) 1962–1967.
40. Tiwari, A., Makhesana, M., Patel, K.M., Mawandiya, B.K. Experimental investigations on the applicability of solid lubricants in processing of AISI 4140 steel. *Material Today: Proceeding* 26(Part 2) (2020) 2921–2925.
41. Zheng, G., Ding, T., Huang, Y., Zheng, L., Ren, T. Fatty acid based phosphite ionic liquids as multifunctional lubricant additives in mineral oil and refined vegetable oil. *Tribology International* 123 (July 2018) 316–324.
42. Wu, X., Yue, B., Su, Y., Wang, Q., Huang, Q., Wang, Q., Cai, H. Pollution characteristics of polycyclic aromatic hydrocarbons in common used mineral oils and their transformation during oil regeneration. *Journal of Environmental Sciences* 56 (June 2017) 247–253.

43. Santos, J.C.O., Santos, I.M.G., Souza, A.G. Thermal degradation of synthetic lubricating oils: Part III – TG and DSC studies. *Petroleum Science and Technology* 35(6) (2017) 540–547.

44. Panchal, T.M., Patel, A., Chauhan, D.D., Thomas, M., Patel, J.V. A methodological review on bio-lubricants from vegetable oil based resources. *Renewable and Sustainable Energy Reviews* 70 (April 2017) 65–70.

45. Nagendramma, P., Kaul, S. Development of ecofriendly/biodegradable lubricants: An overview. *Renewable and Sustainable Energy Reviews* 16(1) (January 2012) 764–774

46. Zareh-Desaria, B., Davoodi, B. Assessing the lubrication performance of vegetable oil-based nano-lubricants for environmentally conscious metal forming processes. *Journal of Cleaner Production* 135 (1 November 2016) 1198–1209.

47. Nurula, M.A., Syahrullail, S. Evaluation between existing and alternative lubricant in metal forming process. *Procedia Manufacturing* 2 (2015) 470–475. 2nd International Materials, Industrial, and Manufacturing Engineering Conference, MIMEC2015, 4–6 February 2015, Bali Indonesia Lubricant viscosity.

48. Aiman, Y., Syahrullail, S. Development of palm oil blended with semi synthetic oil as a lubricant using four-ball tester. *Jurnal Tribologi* 13 (2017) 1–20.

49. Brinksmeier, E., Meyer, D., Huesmann-Cordes, A.G., Herrmann, C. Metalworking fluids—Mechanisms and performance. *CIRP Annals* 64(2) (2015) 605–628.

50. https://www.powermag.com/all-about-lubricant-additives/

51. Tertuliano, I., Figueiredo, T., Machado, G., Cousseau, T., Sinatora, A., Machado, I. The effects of surface conditioning and gear oil type on friction and wear behavior under sliding condition. *Proceedings of the Institution of Mechanical Engineers, Part J: Journal of Engineering Tribology* 232(1) (2017) 73–84.

52. Papay, A.G. Antiwear and extreme-pressure additives in lubricants. *Lubrication Science* 10(3) (1998) 209–224.

53. https://www.ispatguru.com/role-of-lubrication-during-the-process-of-metal-working/

54. Munther, P., Lenard, J.G. Tribology during hot, flat rolling of steels. *CIRP Annals* 44(1) (1995) 213–216.

55. B. Podgornik, V. Leskovek. Wear mechanisms and surface engineering of forming tools. *Materials and Technology* 49(3) (2015) 313–324.

56. Hardell, J. (2007). High Temperature Tribology of High Strength Boron Steel and Tool Steels. Luleå University of Technology, Department of Applied Physics and Mechanical Engineering, Division of Machine Elements, PhD thesis.

57. Tribology in metal forming at elevated temperatures. Available from: https://www.researchgate.net/publication/273888587_Tribology_in_metal_forming_at_elevated_temperatures [accessed Jul 27 2020].

58. Glascott, J., Stott, F.H., Wood, G.C. The effectiveness of oxides in reducing sliding wear of alloys. *Oxidation of Metals* 24(3–4) (1985) 99–114.

59. Chang, Y.N., Wei, F.I. High temperature oxidation of low alloy steels. *Journal of Materials Science* 24(1) (1989) 14–22.

60. Beynon, J.H. Tribology of hot metal forming. *Tribology International* 31(1–3) (January 1998) 73–77.

61. Persson, A., Hogmark, S., Bergström, J. Thermal fatigue cracking of surface engineered hot work tool steels. *Surface and Coatings Technology* 191(2–3) (2005) 216–227. doi:10.1016/j.surfcoat.2004.04.053.

62. Podgornik, B., Hogmark, S. Surface modification to improve friction and galling properties of forming tools. *Journal of Materials Processing Technology* 174(1–3) (2006) 334–341.

63. Özgür A.E., Yalçın, B., Koru, M. Investigation of the wear performance and thermal diffusivity properties of M41 tools steel coated with various film coatings. *Mater Design* 30(2) (2009) 414–417.

64. Tomala1, T.A., Rodriguez Ripoll, M., Badischool, E. Tool – solid lubricant – Workpiece interactions in high temperatures applications. *Procedia Engineering* 68 (2013) 626–633. The Malaysian International Tribology Conference 2013, MITC2013.
65. Hawryluk, M., Ziemba, J. Lubrication in hot die forging processes. *Proceedings of the Institution of Mechanical Engineers, Part J: Journal of Engineering Tribology* 233(5) (2019) 663–675.
66. Uda, K., Azushima, A., Yanagida, A. Development of new lubricants for hot stamping of Al-coated 22MnB5 steel. *Journal of Materials Processing Technology* 228 (February 2016) 112–116.
67. Li L.X., Peng D.S., Liu J.A., Liu Z.Q. An experiment study of the lubrication behavior of graphite in hot compression tests of Ti–6Al–4V alloy. *Journal of Materials Processing Technology* 112(1) (2001) 1–5.
68. Matsumoto, R, Osakada, K. Lubrication and friction of magnesium alloys in warm forging. *CIRP Journal of Manufacturing Science and Technology* 51(1) (2002) 223–226.
69. Buchner, B., Maderthoner, G., Buchmayr, B. Characterisation of different lubricants concerning the friction coefficient in forging of AA2618. *Journal of Materials Processing Technology* 198(1–3) (3 March 2008) 41–47.
70. Wang, L., He, Y., Zhou, J., Duszczyk, J. Effect of temperature on the frictional behaviour of an aluminium alloysliding against steel during ball-on-disc tests. *Tribology International* 43 (2010) 299–306.
71. Wang, F, Lenard, J.G. An experimental study of interfacial friction-hot ring compression. *Journal of Engineering Materials and Technology* 114(1) (1992) 13–18.
72. Behrens, B.A., Meijer, A., Stangier, D., Hübner, S., Biermann, D., Tillmann, W., Rosenbusch, D., Müller, P. Static and oscillation superimposed ring compression tests with structured and coated tools for sheet-bulk metal forming. *Journal of Manufacturing Processes* 55 (July 2020) 78–86.
73. Wang, L., Zhou, J., Duszczyk, J., Katgerman, L., Friction in aluminium extrusion—Part 1: A review of friction testing techniques for aluminium extrusion. *Tribology International* 56 (December 2012) 89–98
74. Sofuoglu, H., Rasty, J. On the measurement of friction coefficient utilizing the ring compression test. *Tribology International* 32 (1999) 327–335.
75. Björk, T., Bergström, J., Hogmark, S. Tribological simulation of aluminium hot extrusion. *Wear* 224(2) (February 1999) 216–225.

11 Electroless Coating Technique, Properties, and Applications

Sulaxna Sharma
THDC Institute of Hydropower Engineering and Technology

Kuldeep Kumar
CCS University

A. Ansari and Munna Ram
Graphic Era deemed to be University

Gyan Singh
Gochar Mahavidhalya

Awanish Kumar Sharma
Graphic Era deemed to be University

CONTENTS

DOI: 10.1201/9781003097082-11

11.1 INTRODUCTION

Surface coatings can be applied to a variety of surfaces and could provide an excellent way to achieve smoother surfaces with enhanced surface properties for a variety of applications including, corrosion resistance, antifouling, antibacterial and tribological properties, etc. Although applying coatings on different types of surfaces such as metal, alloys, wood, glass, etc., might have resulted in a higher expenditure, but due to their remarkable savings in maintenance cost, minimizing the risk associated with equipment failure due to corrosion and enhanced service life of the equipment, etc., these are thought to be an additional viable in extended terms [1]. An additional dimension has been added to the surface coatings in terms of nanomaterials coating that has proven an efficient method to tailor the properties of coatings and consequently enhance the corrosion-resistant, tribological and other physical properties [1,2]. Nanomaterials may be described as having at least one of their morphological features like particles size, grains size, structures size, etc., in the nanosized range i.e., less than 100 nm, and these can be categorized as of zero-dimensional (nanoparticles), one-dimensional (nanotubes, nanowires, and nanorods), or two-dimensional (nanoplatelets, nanosheets, and nanofilms) nanomaterials [1–3]. Nanomaterials also acquire enhanced physical, chemical, magnetic, electronic, thermal, mechanical, and optical properties, which are principally due to their small sizes and higher surface area [3]. The combination of two or more immiscible phases gives rise to composite coatings [4]. The nanocomposite coatings are individuals in which at least one of the ingredient phases has a measurement lesser than 100 nm in size. Nanocomposite coatings are among the most fascinating and fastest-growing areas of research and have attracted great attention in advanced research all over the world due to their different functionality, opportunities to realize exclusive grouping of properties that are unattainable through conventional resources [2–4]. The utilization of nanoparticles into the matrix arrangement has turn out to be a topic of curiosity in chemical along with engineering applications because of impending changes in physical properties of nanocomposites [2,3,5]. These alterations in properties are approached from two features of nanoparticles; the first one is an enlarged surface area and the second one is quantum effects associated with nano-dimensional particle configuration [2,6,7]. These features can alter or improved properties such as strength, toughness, reactivity, magnetic addition to dielectric properties, etc. [2]. The nanocoating technology has significantly influenced the paint industry with an accretion of properties such as self-cleaning, self-healing, high wear, and scuff resistance and has capability to replace toxic chromium coatings [3]. On the other hand, smart nanocoatings have significantly benefited in reducing biofouling and corrosion effects, and these are also developed to counter external stimulation such as heat, pH, humidity, stress, electromagnetic radiation, coating buckle, etc., by liberation of a controlled quantity of inhibitors and can cure and repair damages and defects [1,2,5]. The nanocoatings possess some other remarkable properties, which are very useful for human beings and that is why they are employed in day-by-day practice such as cell phones, garments, high-quality adherence, optical lucidity, photovoltaic, anti-fogging, computers, eye-glasses, etc. [3,8,9]. In the construction field, nanocoatings are used in walls, paints, tiles, windows, floor covering, air filters and are useful as flame-retardant, self-cleaning, wear and scratch resistance, anti-graffiti, anti-fouling, [8,9], etc. In the biomedical field, these are used for engraving protection, surface coverage, drug delivery, and biocompatibility [8,9]. The nanocoating grasps several other important fields such as military, space, nuclear, petrochemical, paper, automobile, semi-conductor industry, [3,8,9] etc. It has been reported that nanocoatings can be deposited on different substrate surfaces by three common deposition methods named mechanical, physical, and chemical deposition. Among these methods, mechanical deposition method is the cheapest one, and it can be accomplished through paint, spray, spin, or dip coating procedures. On the other hand, physical deposition method can be done by sputtering, bonding, or condensation procedures. Finally, chemical bonding methods are often cheaper but require expensive precursors, such as in sol-gel, atomic layer deposition (ALD), and Langmuir techniques [1,3,7]. Each of the aforesaid techniques of skinny film deposition on the surface of the substrate impinge on the consistency along with surface properties such as strength, rupture toughness, and ductility [4]. Each deposition method has its individual pros and cons and to select the best technique all the processing essentials should be studied carefully. For example, conventional coatings should be used with the optimized conditions so that crack-free surfaces, consistency, smoothness and sticking together, etc. can be achieved. Further, the wet coating is economical and can coat multifarious shapes, but on the other side, it might go through from thermal expansion disparity and necessitate the lofty sintering temperatures [4,8,9]. Similarly, the pulsed laser deposition (PLD) and sizzling pressing methods can fabricate impenetrable and homogeneous coatings but unfortunately have the same drawback as dip coating [4]. The sol-gel is a favorite coating method, as it has short extreme temperatures and is comparatively cheaper than other coatings, although it needs costly unprocessed materials as well as a controlled processing environment [4]. Several most frequently exploited inorganic metallics (Mg, Ti, Zn, Cu, Zr, etc.) are capable of enduring enormous heat and harsh process conditions. From a variety of nano-composite coating technologies, electroless coating technology hold out with fresh advancement in nanoscience and technology to apply coatings over a variety of surfaces for different applications. For the past few decades, several researchers have anticipated diverse schemes to manage different structures like pH, types and concentration of second-phase particles, temperature, etc., for electroless Ni-P-based composite coatings effectively [2,5]. The co-deposition of second-phase nanoparticles into electroless Ni-P medium leads this classical electroless coating technology to an additional pace that is known as electroless nano-composite coatings. It has been lately accounted that Ni-P electroless micro/nano-composite coatings may play a vital role to avoid corrosion and enhance wear and hardness resistance in the case of metals as well [2,5,10].

11.2 ELECTROLESS COATING TECHNOLOGY

Electroless Ni plating was accidentally discovered by Brenner and Riddell in 1946 [11] when they observed that the additive NaH_2PO_2 caused apparent cathode efficiencies of more than 100% in a nickel electroplating bath. They concluded that some chemical reduction was involved in this process and named *electrodeless* plating. Later, the advances in research developed the original process and soon the name lost *'de'*; therefore, the process is known as electroless coating, however, 'autocatalytic' term is also used for the process. The name electroless is somewhat misleading as no external electrodes are present, but charge transfer (i.e., electric current) is involved in the process. In this process, the metal is supplied by the metal salt instead of using an anode, and replenishment is achieved by further addition of metal salt. In the process, the cathode is not used for the reduction of metal; however, a reducing agent is used to provide the electrons; i.e., serving as a cathode, since no external current is used in this process; that is why, it is named as electroless [12–14].

Moreover, electroless coating is among the simple and best coating technologies, in which deposition is a purely systematic chemical process, in which the pretreatment procedure used prior to plating consisted of the following steps (Figure 11.1): (1) cleaning, (2) surface modification, (3) sensitization, (the sensitization in addition to catalyzing stands for deposition of Sn(II)/Sn(IV) ions as of acidic SnCl$_2$ solution), and (4) activation (substrate activation done by dipping in PdCl$_2$ solution); rinsing of the substrate with deionized water is necessary in between the steps [2,5,10–14].

The purpose of sensitizing ions is that it will reduce lively metal (see Equation 11.1).

$$Pd^{2+} + Sn^{2+} \rightarrow Sn^{4+} + Pd^0 \tag{11.1}$$

After the pretreatment, plating on top of the sensitized surface takes place according to Figure 11.2. The solutions of SnCl$_2$, PdCl$_2$ and the solutions of the electroless bath before and after the reaction are shown in Figure 11.3.

11.3   PROCESS MECHANISM FOR ELECTROLESS COATINGS

Two different types of mechanisms have been proposed following the pretreatment process that are discussed in the following sections:

(a) **An electrochemical mechanism**: Here it is exemplifying that catalytic decomposition of hypophosphite goes along with negatively charged electrons which on catalytic face decrease Ni as well as H ions according to the Equations (11.2–11.5) [2,12–15].

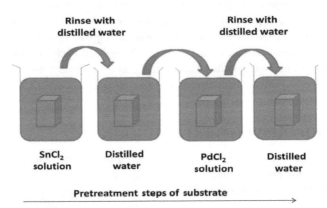

Rinse with distilled water   **Rinse with distilled water**

SnCl$_2$ solution   Distilled water   PdCl$_2$ solution   Distilled water

Pretreatment steps of substrate

FIGURE 11.1   Pretreatment steps of the substrate surface.

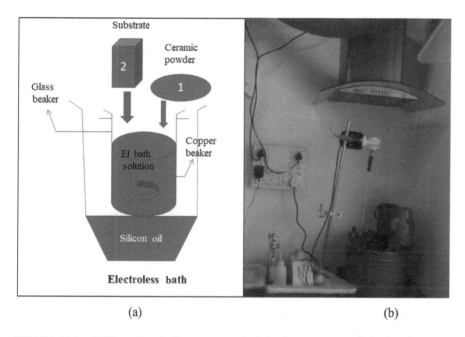

(a)   (b)

FIGURE 11.2   (a) Electroless bath components and (b) picture of electroless bath setup.

(a)   (b)

FIGURE 11.3   The picture of (a) pretreatment solutions and (b) electroless bath solution before and after coating.

$$H_2PO_2^- + H_2O \rightarrow H_2PO_3^- + 2H^+ + 2e^- \qquad (11.2)$$

$$Ni^{+2} + 2e^- \rightarrow Ni \qquad (11.3)$$

$$2H^+ + 2e^- \rightarrow H_2 \qquad (11.4)$$

$$H_2PO_2^- + 2H^+ + e^- \rightarrow P + 2H_2O \qquad (11.5)$$

(b) **An atomic hydrogen mechanism**: Here it is exemplifying that H is liberated owing to de-hydrogenation of hypophosphite compound adsorbed on exterior face and following Equations (11.6–11.9) involved in the process [2,5,12–15].

$$H_2PO_2^- + H_2O \rightarrow HPO_3^{2-} + H^+ + 2H_{ads} \tag{11.6}$$

$$2H_{ads} + Ni^{+2} \rightarrow Ni + 2H^+ \tag{11.7}$$

$$H_2PO_2^- + H_{ads} \rightarrow H_2O + OH^- + P \tag{11.8}$$

$$H_2PO_2^- + H_2O \rightarrow H^+ + HPO_3^{2-} + H_2 \tag{11.9}$$

This is contingent to facilitate catalytic layer where hypophosphite is concentrated through H_{ads} as well as P. Here enlarge commencement of nano-sized alloy captivating catalytic location acquires position first, horizontally after then vertically [14]. The plating rate is more in horizontal than vertical route which makes deposition pore liberated.

11.4 ROLE OF INDIVIDUAL COMPONENTS IN THE ELECTROLESS BATH

In the electroless coating technique, the chemicals used in the coating process play a vital role for the smooth and effective surface coating a few of them have been discussed in the following section and summarized in Table 11.1.

TABLE 11.1
Components of Electroless Bath and Their Functions [2,18]

Components	Function
Source of metal ion	$NiCl_2$, $NiSO_4$ and $(CH_3COO)_2Ni$
Reducing agent	Sodium hypophosphite
Complexing Agent	Form nickel complexes to control free Ni ion in bath solution to prevent precipitation of Ni phosphate and serves as pH buffers. For example, mono-/ dicarboxylicacids, hydroxyl-carboxylic acids, ammonia, alkanolamines, etc.
Accelerators	It accelerates deposition process by activating the reducing agents, in other words it has opposite action to the stabilizers and complexing agents, e.g., anions of some mono- and di-carboxylic acids, fluorides borates, etc.
Stabilizer (inhibitors)	Stabilizers provide shielding to the catalytically active nuclei and prevent the decomposition of bath solution. A few are as follows: Pb, Sn, As, Mo, Cd or Th ions and thiourea, etc.
Buffers	To maintain pH of solution
pH regulators	pH regulators adjust the pH subsequent to reaction, e.g., sulfuric acid, hydrochloric acids, soda, caustic soda, and ammonia solution
Wetting agents	Used to increase the wettability of active surfaces to be coated

11.4.1 SOURCE OF NICKEL IONS

In the electroless bath, nickel sulfate ($NiSO_4$) and nickel chloride ($NiCl_2$) are mainly used as a source of Ni^{2+} ions for depositing amorphous alloys. It has been reported [2,16,17] that concentration variation of Ni^{2+} ions in electroless bath solution has very little influence on the reduction rate of nickel ions. However, high concentrations of Ni^{2+} ions can cause the formation of rough deposits and thus deteriorates the quality of coatings. However, sulfate and chloride salts of nickel result in systematic and arbitrary [16–17] depositions.

11.4.2 REDUCING AGENTS

Hypophosphite, borohydride/ dialkyl-amino borane and hydrazine are mainly used reducing agents. Based on the use of these reducing agents the coatings are coupled as (Ni-P) nickel phosphorous, (Ni-B) nickel boron as well as a pure nickel (Ni) in the electroless coating baths [19]. Hypophosphite is the well-built and inexpensive reducing agent for electroless Ni-P coating as if these are compared with Ni-B electroless coatings [2,5,19–20].

11.4.3 COMPLEXING AGENTS

In the electroless coating technique, complexing agents play an important role to maintain the bath composition. As the reaction progress, nickel ions quantity decreases continuously. The bath cannot be stock up due to excessive precipitation of nickel phosphate that results solution decomposition and a rough dark coating occurs with deteriorated surface quality. The complexing agents prevent the uncontrolled reduction of nickel ions into solution to some extent owing to the formation of complexes by nickel ions. In the literature, several complexing agents have been reported [5], e.g., glycolic/succinic/malonic acids, sodium citrate, ethylenediamine, etc. It has been reported [2,5,16] that for sodium citrate best results occur at the concentration 30 g/l which effectively reduces the creation of nickel phosphite and helps to check coating to become dull and porous. In addition, some stabilizers e.g., Bi, Pb, Cd, Te, thiourea, and cyanides in the quantity of little (ppm) parts per million can help in stabilizing the electroless bath [5,16].

11.4.4 ACCELERATOR

As complexing chemicals decrease deposition pace, therefore, to pick up the deposition pace some salts e.g., carbonic acids or soluble fluorides, etc., can also be added to the bath. These accelerators are also economical. The main function of the accelerator is to break the H-P bond of the hypophosphite molecule which results in easy removal of phosphorous and adsorption at the catalytic face. Succinic acid is the most frequently used accelerator in hypophosphite-reduced solutions [2,5].

11.4.5 OTHER PROCESS PARAMETER

The operating temperature of the bath is an important parameter and it can affect deposition rate. The plating rate below 65°C temperature is low, and it is more as temperature increases and correct for all plating processes. Bath operational temperature is roughly 90°C and exceeding this aqueous solution starts evaporating and the bath tends to become unstable [2,20]. Further properties of substrate surface also affect the deposition process e.g., higher magnetization of deposits has been exhibited on aluminum substrates than on a brass substrate. The pH of bath needs to be attuned properly because it is an essential parameter and can affect phosphorus content in deposits. If the alkalinity of the solution increases, there is a decrease in phosphorus content in deposits. One can have low, medium, and high phosphorus content deposited by adjusting pH of the bath (See Table 11.2). The acidic bath produces an extremely intense and even surface of nickel deposit. However, alkaline bath generates only a dazzling surface.The bath loading factor suggests that the deposition rate increases as the bath loading factor increases. There is a decisive value beyond which the bath decomposes entirely [2,10,19,20].

Heat treatment is an important and essential part of the electroless coating process. It can improve the adhesion of deposits to the substrate by diffusion and improves mechanical properties of the coatings e.g., hardness, structure, morphology of deposits [5,10,19] and transformation of deposits from amorphous to crystalline phases. Aggarwala et al. [2,19,21] calculated the hardness of electroless Ni-P deposits as annealing heated and this group concluded that the optimum heat treatment temperature for electroless Ni-P deposits is around 400°C/h. At this temperature, the deposits get crystallized as fine particles of the Ni_3P phase, which is evidenced by the powder XRD technique and attributed to the increased hardness of the deposits. Heat treatment at higher temperature for a longer duration can cause the growth of nickel grains and phosphide coarsening, which leads to a decrease in progressive hardness. Apart from heat treatment temperature, hardness is also affected by the time of heat treatment and the percentage of phosphorous content in the deposit. For example, to harden the surface, low phosphorous (~3%) and high phosphorus (10.5%–12%) deposits the recommended temperature for heat treatment is ~350°C and 400°C [2,10,18,19,21] respectively. Similarly, for electroless nickel boron (Ni-B) deposits, annealing temperature is testimony at 350°C and 450°C for 1h, which transforms the amorphous phase of deposit to crystalline Ni_3B and Ni_2B phases. Beyond temperature 450°C, the augmentation of crystalline Ni and adaptation of Ni_2B phase

---

TABLE 11.2

Phosphorus Content of Ni-P and Ni-B Alloy Coatings in Acidic Bath [2,5,19,21]

S. No.	Ni-P Alloy (% P)	Ni-B Alloy (% B)	Coating Properties
1	3–5 (L)	0.1–2 (L)	Excellent wear resistance and corrosion resistance in concentrated caustic soda.
2	6–9 (M)	2–5 (M)	Shows corrosion protection and abrasion resistance. The plating bath works economically.
3	10–14 (H)	5–10 (H)	Very ductile and corrosion resistant against chlorides solutions.

L=Low; M=Medium; H=High.

to a more stable Ni$_3$B phase [20] occur. This improved adherence and strength of the coatings after heat treatment is whispered to come beginning in chemical surface alteration by dissemination bonding between substrate metal (e.g. Al, Cu, Steel, etc.) and the alloy coating [22]. This diffusion mechanism is not applicable for ceramics and glass substrates. Also, corrosion resistance properties of electroless nickel deposition have been affected by heat treatment. It has been instituted that as-plated coatings show evidence of better corrosion-resistant properties as compared to heat-treated coatings due to the nebulous character of as-plated depositions. It is discussed in previous paragraphs, crystalline nature of plating increases as heat-treatment is increased which results in an enhancement in the grain boundaries that serve as vigorous sites for corrosion attack [2,5,19]. In a topical report [23] a novel sort of thermo-chemical action has been proposed, which can institute to turn out enhanced corrosion battle in EL Ni–B platting as put side by side to normal annealing healing.

11.5 TYPES OF ELECTROLESS COATINGS

Electroless coatings are broadly grouped into three major types, as follows:

11.5.1 Metallic Coatings

Pure electroless nickel is of great use mainly for semiconductor applications. This is produced by using hydrazine as a reducing agent for nickel ions present in the bath. However, the industrial interest of the coating is restricted due to its high cost and hazardous nature. However, electroless copper precoating is being used for electroplating on plastics, ceramics, polymers, and non-conducting materials [2,21]. In shielding electronic devices, electroless copper coating is also utilized. The thickness of these coatings varies from 0.12 to 3 μm. The typical nickel plating bath composition is as follow: nickel acetate (60 g/l), glycolic acid (60 g/l), EDTA (ethylene diamine tetrasodium salt, 25 g/l), hydrazine (100 g/l), sodium hydroxide (30 g/l), pH 10.5–11, and temperature 85°C–90°C [2,5,19,21]. Other types of electroless metallic coatings have been reported by Hajdu [22] in detail.

11.5.2 Electroless Nickel Alloy Coatings (e.g., Ni-P and Ni-B Alloys Deposits)

Several binary, ternary, and quaternary alloys coatings e.g., Ni-P and Ni-B, Ni-P-B and Ni-W-P, etc., have been deposited and reported in the related context. Nickel is the single widely coated element along with phosphorous and boron. These coatings possess paramagnetic properties and show other many industrial applications such as excellent resistance to wear and corrosion and high hardness. For the deposition of Ni-P and Ni-B alloy coatings, two types of baths, namely, acidic and basic baths, have been employed.

a. **Acidic bath for Ni-P and Ni-B**: A typical acid bath composition reported for Ni-P (nickel sulfate: 33 g/l, sodium hypophosphite: 20 g/l, lactic acid: 28 g/l, sodium succinate: 16 g/l, lead: 0.003 g/l at pH 5–6 and temperature

85°C–90°C)and Ni-B (nickel chloride: 30 g/l, di-ethylamine borane: 3 g/l, methanol: 40 g/l, di-methylamine borane (DMAB), 4 g/l, sodium acetate: 20 g/l, sodium succinate: 20 g/l and sodium citrate: 10 g/l, pH 5–6 and temperature 50°C–60°C) [2,5,19,21]. For Ni-B coatings, DMAB is used as a reducing agent with varied boron content (0.1%–4%). Ni-B coatings have higher hardness as compared to Ni-P coatings and, therefore, used very frequently for industrial wear applications.

b. **Alkaline bath for Ni-P and Ni-B**: To produce 10 μm/h rate of deposition the hot alkaline bath composition of Ni-P is (nickel chloride: 30 g/l, sodium hypophosphite: 10 g/l, ammonium citrate: 65 g/l, ammonium chloride: 50 g/l, pH 8–10 and temperature ~ 80°C–90°C) above 90°C the pH of the bath decreases suddenly due to loss of ammonia and causes high instability of the bath which is the main disadvantage of alkaline bath [2,5,21]. However, hot and cold alkaline baths are used for Ni-B and the bath composition for hot condition is (nickel chloride: 30 g/l, ethylenediamine: 60 g/l, sodium borohydride: 1.2 g/l, thallium nitrate: 0.07 g/l, NaOH: 40 g/l, pH: 14 and temperature: 90°C) the composition reported to have deposition rate of 20–25 μm/h. The composition of cold alkaline bath is (nickel sulfate: 30 g/l, DMAB: 3 g/l, ammonium citrate: 15 g/l, NH$_4$Cl: 15 g/l, 2-metcaptobenzothiazole: 0.0002 g/l, pH ~ 7.5 and temperature 25°C–35°C) the composition is reported to have deposition rate is 7–12 μm/h [2,5].

It has been reported that low-temperature alkaline baths are used to deposit nickel on plastics as well as may provide good solderability for the electronic industry. However, lower adhesion to steel, poor corrosion resistance, and difficulty in processing aluminum at high pH, are the limitations. The thickness of deposition is very much temperature dependent and so the alkaline bath found little industrial use. To plate steel and other metals, hot acid hypophosphite-reduced baths are frequently used. Whereas, for plating plastics and non-metals [21] warm alkaline hypophosphite baths are used. Several other factors can affect the bath life such as surfactants, stabilizers, bath temperature, etc., [2,5,6,19,21]. In this way, hot acid bath has several advantages over alkaline baths including the high thickness of better-quality deposits onto a metal surface and relatively high stability of bath solution during plating. Ternary and quaternary alloy coatings are the extension of binary coatings and are discussed elsewhere [6,24].

11.6   ELECTROLESS NICKEL COMPOSITE COATINGS

Electroless composite coatings are represented by dispersed second-phase particulate material into a metal matrix. Since the co-deposition of second-phase particulate material including (1) hard (diamond, SiC, and Al$_2$O$_3$, etc.) and (2) soft (PTFE, MoS$_2$, graphite, BN, etc.) particles can further enhance hardness, anti-sticking, lubricity, and anti-wear properties [2,5,21,25]. Depending on the requirement, soft and hard particles are added as second-phase particles (i.e., reinforcement) to the metal matrix. For wear-resistant applications, hard particles such as SiC, WC, Al$_2$O$_3$, etc., are added, whereas to reduce the friction coefficient the lubricating solid particles like MoS_2, graphite, etc., are added [2,5,19,21,25]. Thus, these types of composite coatings show a wide spectrum of applications e.g., claddings, anti-reflection films, process machinery and electronics industries, etc. [9,17,25].

In metallic matrix, the co-deposition of Al_2O_3 and PVC particles was first studied in 1966 [26]. Thereafter, several research studies have been carried out research in this direction all over the world and a few of them are discussed here. Sharma [27] and Metzger and Florian [28] have introduced alumina (Al_2O_3) particles of micron size into electroless nickel coatings. In Wankel (rotary) internal combustion engine, Ni-P-SiC coatings were used commercially for the first time [26]. The methodology employed to produce composite coatings was like conventional electroplating. However, dispersion of finely divided second-phase particles in the plating bath substantially increases the surface area of loading up to 800 times that leads to homogenous decomposition of the bath [29]. Thereafter, improved and novel methods have been reported for the coating procedure including proper mixing, addition of surface-active substance, *in-situ* generation of second-phase particles nanoparticles [30], etc. Two different types of mechanisms for particle incorporation in electroless Ni-P matrix have been reported. The first one is *conventional method* in which the externally synthesized second-phase particles have been added to the electroless bath during the deposition process in this way the second-phase particles co-deposited along with Ni-P. The second one is a non-conventional method; in this method, by adding suitable precursors to the electroless bath, second-phase particles have been synthesized inside the bath (i.e., *in-situ* generation) and get dispersed simultaneously into the Ni-P matrix. By these methods, several second-phase particles such as Al_2O_3, BN, SiC, TiO_2, B_4C, MoS_2,C (graphite/diamond), PTFE and Si_3N_4, etc., [2,5,19,21,25–49] of micron-size have been co-deposited successfully into electroless Ni-P matrix.

The effective co-deposition of second-phase particles into Ni-P matrix depends on various factors including particle shape, size, relative density and charge on particle, catalytic or inert nature of particles, and concentration of particles in the plating bath. Apart from this, electroless bath composition, method and degree of agitation, compatibility of the particle with the matrix, and the orientation of the particle being plated have also affected the process of deposition and coating properties. During the deposition process, the **convection forces** also play an important role, which moves the particles straight to the substrate surface before being incorporated into the growing coating. The **size of the particles** has a definite impact on their incorporation in the electroless Ni-P matrix. In general, it is recommended that particles size should be selected according to the thickness of the electroless nickel deposit. **Particle shape** also plays a vital role in determining their incorporation level. From the previous studies, it has been concluded that angular particles have a greater tendency to incorporate into metal matrix than round ones. However, better results have also been obtained with spherical particles of alumina. The difference in particle shape also affects the finishing of the deposit. Very smooth and very rough surfaces were obtained from small rounded particles and large angular particles, respectively [2,5,18,20–22,24,27–29]. The possibility of second-phase particles to get agglomerate cannot be ruled out beyond a critical **concentration**. Because at high concentration, the mean distance between the particles is decreased, resulting in a decrease in the level of incorporation settlement of the particles. Therefore, the concentration of second-phase particles within the bath is very important in the distribution of particles in the deposit [30,31]. *Surfactants* have also played an important role in the incorporation of second-phase particles especially for soft and lubricating particles, e.g., polytetrafluoroethylene (PTFE), graphite, and molybdenum disulfide [32–35].

11.7 INCLUSION OF SECOND-PHASE (X) PARTICLES INTO Ni-P MATRIX

The widespread investigations were done to insert second stage nanoparticles into EL Ni-P medium [18,20,21,27,29–35]. Lanzoni et al. [36] described assimilation of SiO_2 nanoparticles in EL Ni–P plating extensively improve micro-hardness (heat treatment $\sim 400°C$) and wear conflict of plating. Corrosion fighting quality of EL Ni-P-SiO_2 platting beside salty solutions has appreciably enhanced and SiO_2 has (2 wt%; size ~ 20 nm) [37,38] characteristics. To build up Ni-P-TiO_2 composite plating a novel method, i.e., the addition of transparent TiO_2 solution in electroless is developed [30,39,40]. The TiO_2 powder having ~ 10 nm size has also been coated with Ni-P [39] and results improved triobological properties. Micro-hardness test values demonstrate a noteworthy improvement from ~ 710 $HV_{0.2}$ to 1025 $HV_{0.2}$. In addition, soft particles (PTFE, Hexa-BN, MoS_2, WS_2, CNT and C) offer superior lubrication when incorporated in EL Ni–P medium [2,5,19,32–35]. The Ni-P-SiC deposition has price effectual plus best-performing confederacy [41–43] for scratchy and wear conflict properties. The stripped of Ni-P-X (X= hard particles), Ni-P-SiC depositions demonstrate high hardness along with wear resistance to replacing active "stiff chromium" coating into aerospace engineering [41–43]. Along with others, the grouping (Ni-P-SiC coating) has established a lucrative (price tag wise) as well as best-performing arrangement functioned in industry. Alumina powder (Al_2O_3 of size 50, 300, and 1000 nm) is co-deposited into Ni-P medium and studied [27] the outcome of particle size on hardness, microstructure in addition to corrosion protection properties. The barium hexaferrite ($BaZn_2\text{-}yCoyFe_{16}O_{27}$) is a central magnetic material and is extensively utilized as magnetic recording media, enduring magnet and microwave absorbers [44]. Therefore, attempts to incorporate nano-barium hexaferrite into electroless Ni-P matrix have also been made and reported to show improved microwave absorption properties [44]. Increased wear resistance property has been observed for Ni-P-X particles (X= $Cr_3C_2$27% volume, teflon, diamond) [2,5,45,46]. The WinowlinJappes et al. [45] has studied corrosion protection of Ni-P-diamond composite plating against NaCl (3.5 wt%) solution and reveals subsidiary raise in corrosion conflict contrast to plain Ni-P plating. The electrochemical resistance of composite coatings Ni-P-CeO_2 [47] and Ni-P-TiO_2 [30] is determined by the EIS method. The superior corrosion protection is observed here in contrast to plain Ni–P deposits because of the smaller tendency of CeO_2-containing coating undergo to local-cell corrosion. The Ni-P-Fe_3O_4 composite coatings exhibit superior resistance against corrosion in contrast to plain Ni–P plating and show superior protection in the air against cyclic test at temperature 800°C [48]. Shibli et al. [49] reported improved corrosion resistance on the assimilation of ZnO nanoparticles into Ni–P matrix and Sharma et al. [18] reported Ni-P-ZnO for antibacterial and anticorrosion applications. Corrosion and tarnish protection properties of Ni-P-X (where X = SiO_2, CeO_2, $Zn_3(PO_4)_2$, $ZnSnO_3$, and $ZnSiO_3$) [37,38,47,50] nanoparticles have also been discussed. The mechanism for co-deposition of nano-meter diamond particle along with Ni–P has been explained [47] and studied for their anti-fretting and anti-wear properties. It has been observed that it shows brilliant anti-fretting and wear properties for elevated load at normal temperature than nano-diamond composite plating arranged by electro-brush plating [45]. Higher micro-hardness of these coatings especially at relatively high temperature is observed as compared to conventional Ni-P coating. Increased oxidation resistance for Ni-P-X (where X = Si_3N_4 (8.01 wt%), CeO_2 (7.44 wt%) or TiO_2 (5.42 wt%)) has been reported [51]. Bozzini et al. [52] described the magnetic dependency of Ni–P–B_4C in as-plated and annealed conditions. In as-plated form, the coating exhibits a strapping reliance of magnetic susceptibility in a functional magnetic field because of better structural heterogeneity in contrast to plain Ni–P coatings. Further on heat treatment, composite coatings sealed field reliance with superparamagnetic character, whereas plain Ni-P deposits develop into ferromagnetic nature. The diversity in the magnetic performance of these depositions is due to the concentration of B_4C (25 volume %) which results in nickel-phase precipitation. The CNTs may be permeating into an extremely minute opening in spongy Ni-matrix to construct passive coat additional stable and compressed that is why; it shows tremendous corrosion protection and elevated micro-hardness. The EL Ni-P-CNT composite coatings have spacious applications in contemporary electronics industry [53–55]. The CNFs coated upon Ni–P alloys [56] catalytic surface is evidence for unsystematic orientation, hovering of equivalent graphite planes with imperfections twisted from their axis. It has been reported [56] that Ni-P catalyzed CNFs divulge outstanding turf emanation properties. The CNF is a lightweight, relatively low-cost, high tensile strength, high hardness, and excellent corrosion-resistant material. Ni-P-TiO_2/ZrO_2 coatings have been reported [57,58] for corrosion, hardness, and wear resistance improved than the basic substrate and Ni-P coatings. Furthermore, Ni-P-WO_3 coatings have been reported for their electrocatalytic activity [59]. Ni-P-Ti coatings show indentation and bending behavior [60] and scratch-resistance calculated [61]. Luo et al. [62] have studied Ni-W-P/Ni-P-nano-ZrO_2 coatings for structures, morphologies, and corrosion resistance properties. Dhakal et al. [63] have studied composite Ni-P-TaC coating for microstructural and electrochemical corrosion properties.

11.8   PROPERTIES AND APPLICATIONS OF ELECTROLESS COATINGS

Some of the chemical, physical, electrical and mechanical properties of varying concentrations of phosphorous (low, medium, and high) content are summarized in Table 11.2 in the previous section. In addition to that electroless coatings show high hardness which is an important parameter for many successful applications. However, the hardness of the coatings is strongly influenced by the phosphorus content in as-deposited conditions. Heat treatment of electroless coatings in the optimum temperature range (for Ni-P it is 345°C to 400°C) results in a dramatic increase in the range of ~850 to 950 HK_{100} in hardness. For phosphorus content within the normal operating range [2,5,19,20–22] electroless nickel deposits possess a low ductility. The coatings are hard and brittle, and elongation to fracture is typically 1%–2.5%. On heat treatment, the hardness of electroless nickel increases, whereas ductility reduces. Because of high hardness and good natural lubricity, electroless nickel coatings exhibit good wear resistance. Wear resistance of deposits can be enhanced by heat treatment and co-deposition of particulate matter (e.g., silicon carbide, diamond, alumina, fluorinated carbon, and polytetrafluoroethylene). In the as-plated condition, electroless nickel coatings have good resistance to erosion and abrasion. However, heat treatment of deposits can significantly improve hardness and wear resistance. Therefore, the hardness of electroless nickel deposits depends mainly on heat treatment and phosphorus content that can be advantageous to prevent galling. Low phosphorus deposits provide better wear properties [2,5,19,20–22] in the as-deposited condition, and this would naturally be attributed to greater hardness. The major reason for the widespread use of electroless nickel is due to its high corrosion resistance, which enables it to use as a protective coating. Though all electroless nickel deposits do not perform in the same way however their performance is highly dependent on the specific conditions of the coatings. For example, high phosphorus deposits are a good performer at elevated temperature than low phosphorus deposits, as well as are superior in neutral or acidic media than strongly alkaline media. Electroless nickel deposits have very low porosity and uniform thickness. As the thickness of coating increases porosity decreases, and for deposits, phosphorus (containing ~ 10% P) [19–22] tendency to form porous coatings is greatly reduced. However, on heat treatment, an increase in porosity of the coating generally occurs, which decreases corrosion resistance. This is probably due to the formation of microcracks [2,5,19–22] and is of most concern with high phosphorus deposits. In addition to this electroless coatings possess a low friction coefficient that is useful in self-lubricating coating, as well as act as protective coating owing to its good chemical resistance. These coatings are useful in many functional applications due to their promising properties of solderability and weldability.

Electroless nickel coatings are mainly used for wear resistance, corrosion resistance, lubricity, buildup of worn over machined surfaces, and solderability. By all segments of the industry in varying degrees, these properties are utilized either separately or in combination such as oil field valves, drive shafts, rotors, paper handling equipment, fuel rails, doorknobs, kitchen utensils, and optical surfaces for diamond turning, bathroom fixtures, electrical/mechanical tools, office equipment, etc. It is also commonly used as a coating in electronics printed circuit board and hard disk drive manufacturing. The coating can be used to salvage worn parts due to its high hardness. Its use in the automotive industry for wear resistance has increased significantly.

11.9 SUMMARY

Electroless deposition technique which increases the service life and durability of the components used in various engineering applications, on considering, the cost effectiveness factor. In recent years, most researchers have been focusing on electroless nickel(phosphorus/boron) composite coatings as compared to plain electroless nickel coatings because of their optimistic future in wear-resistant, self-lubricating, anti-corrosion, and other promising properties. Non-conventional methods of electroless composite coatings have also attracted the attention of researchers and there is a scope for further research to optimize the parameters of the in-situ generation of reinforcing material. The electroless Ni-P-X and Ni-B-X (where X= SiC, CeO_2, ZrO_2, TiO_2, $ZnOAl_2O_3$, CNTs, Fe_3O_4, B_4C, diamond, etc.) composite coatings have been developed successfully by conventional method in micron and nano-size range and studied for different properties including wear and corrosion resistant, high hardness, microwave absorption range, magnetic properties, antibacterial properties, lubricating and anti-scratch properties, etc. Owing to the good response observed for abovementioned properties these coatings provide excellent applications in defense, textiles, heavy industries, and several other segments of industry. In addition to the improved properties of the coatings, some features need to be improved for better performance of the electroless composite coatings, for example, non-consistent allocation of nanoparticles on the coated surface that may serve as a source of cavitation and reduces the coating performance.

REFERENCES

1. B. Fotovvati, N. Namdari and A. Dehghanghadikolaei, 'On coating techniques for surface protection: A review', *J. Manuf. Mater. Process.*, 3 (2019) 28.
2. R.C. Agarwala, V. Agarwala and R. Sharma, 'Electroless Ni-P based nanocoating technology—A review', *Synth. React. Inorg., Met.-Org.,Nano-Met. Chem.*, 36(6) (2006) 493–515.
3. J. Jeevanandam, A. Barhoum, Y. S. Chan, A. Dufresne and M. K. Danquah, 'Review on nanoparticles and nanostructured materials: History, sources, toxicity and regulations', *Beilstein J. Nanotechnol.*, 9(2018) 1050–1074.
4. P. Nguyen-Tri, T.A. Nguyen, P. Carriere and C.N. Xuan, 'Nanocomposite coatings: Preparation, characterization, properties, and applications', *Int. J. Corros.*, 2018, Article ID 4749501, 19 pages doi:10.1155/2018/4749501.
5. S. Jothi, M. Rajaraman, T. Tamilarasan, S. Udayakumar, S. Andiappan, Electroless composite coatings, in *'Electroless Nickel Plating: Fundamentals to Applications'*, Edited by F. Delaunois, V. Vitry and L. Bonin, CRS Press, Ed Ist, 2019.
6. E. Gaffet, 'Nanomaterials: A review of the definitions, applications, health effects. How to implement secure development', https://www.researchgate.net/publication/51910938.
7. A. Al-Azzawi1 and P. Baumli, 'Methods of composite coating: A review', *Mater. Sci. Eng.*, 40(1) (2015) 26–32.
8. https://www.nanowerk.com/nanocoatings.php.
9. https://www.futuremarketsinc.com/nanocoatings-buildings/.
10. Y. De Hazan, F. Knies, D. Burnat, T. Graule, Y. Yamada-Pittini, C. Aneziris and M. Kraak, 'Homogeneous functional Ni-P/ceramic nanocomposite coatings via stable dispersions in electroless nickel electrolytes', *J. Colloid Interface Sci.*, 365 (2012) 163–171.
11. A. Brenner and G.E. Riddell, 'Nickel coating on steel by chemical reduction', *J. Res. Natl. Bur. Stand.*, 37 (1946) 31–34.
12. W. Riedel. *'Electroless Nickel Plating'*, Finishing Publications Ltd., Great Britain, 1991.
13. P. Gillespie, Electroless nickel coatings: Case study, in *'Surface Engineering Casebook'*, Edited by J.S. Burnell-Gray and P.K. Datta, Woodhead Publishing limited, 1996, pp. 49–72.
14. N. Feidstein, T. Lancsek, D. Lindsay and L. Salerno, 'Electroless composite Plating', *Met. Fin.*, 81 (1983) 35–41.

15. J.P. Marton and M. Schlesinger, 'The nucleation, growth, and structure of thin Ni-P films', *J. Electrochem. Soc.*, 115(1968) 16–21.

16. J.N. Balaraju, T.S.N. Sankara Narayanan and S.K. Seshadri, 'Electroless Ni–P composite coatings', *J. Appl. Electrochem.*, 33 (2009) 807–816.

17. N. Feldstein, G.O. Mallory and J.B. Hajdu, *'Electroless Plating: Fundamentals and Applications'*, American Electroplaters and Surface Finishers Society, Orlando, FL, 1990, pp. 269.

18. S. Sharma, S. Sharma, A. Sharma and V. Agarwala, 'Co-deposition of synthesised ZnO nano-particles into Ni-P matrix using electroless technique and their corrosion study', *J. Mater. Eng. Perform.*, 25(10) (2016) 4383–4393.

19. K. Krishnaveni, T.S.N. Sankara Narayanan and S.K. Seshadri, 'Electroless Ni–B coatings: Preparation and evaluation of hardness and wear resistance', *Surf. Coat. Technol.*, 190(1) (2005), 115–121.

20. J. Sudagar, J. Lian and W. Sha, 'Electroless nickel, alloy, composite and nano coatings–A critical review', *J. Alloys Compds.*, 571 (2013) 183–204.

21. R.C. Agarwala and V. Agarwala, 'Electroless alloy/composite coatings: A review', *Sadhana*, 28 (2003) 475–493.

22. Hajdu, J., Chapter 7 Surface preparation for electroless nickel plating, In; Hajdu, J., Mallory, G.O., Editors, Electroless Plating: Fundamentals and Applications; American Electroplaters and Surface Finishers Society: Orlando, FL, USA, 1990; pp. 193–206.

23. C.T. Dervos, J. Novakovic and P. Vassiliou, 'Vacuum heat treatment of electroless Ni-B coatings', *Mater. Lett.*, 58 (2004) 619–623.

24. J.N. Balaraju, C. Anandan and K.S. Rajam, 'Morphological study of ternary Ni–Cu–P alloys by atomic force microscopy', *Appl. Surf. Sci.*, 250 (2005) 88–97.

25. P. Sahoo and S.K. Das, 'Tribology of electroless nickel coatings–A review', *Mater. Design.*, 32 (2011) 1760–1775.

26. J.M. Odekerken, 'Method of Electrodepositing a Corrosion Resistant Nickel–Chromium Coating and Product Thereof', US Patent, 282,810,1966.

27. S.B Sharma, 'Synthesis and tribological characterization of Ni-P based electroless composite coatings', PhD Thesis, IIT Roorkee, India, 2002.

28. W. Metzger and T. Florian, 'Electroless nickel plating', *Trans. Inst. Met. Finish.*, 54 (1976) 174.

29. J.N. Balaraju and S.K. Seshadri, 'Preparation and characterization of electroless Ni-P and Ni-P-Si$_3$N$_4$ composite coatings', *Trans. Inst. Met. Finish.*, 77(2) (1999) 84–86.

30. W. Chen, W. Gao and Y. He, 'A novel electroless plating of Ni–P–TiO$_2$ nano-composite coatings', *Surf. Coat. Technol.*, 204 (2010) 2493–2498.

31. Z. A. Hamid and M.T. AbouElkhair, 'Development of electroless nickel–phosphorous composite deposits for wear resistance of 6061 aluminum alloy', *Mater. Lett.*, 57 (2002) 720–726.

32. S.M. Moonir-Vaghefi, A. Saatchi and J. Hejazi, 'Deposition and properties of electroless nickel-phosphorus-molybdenum disulfide composites', *Met. Finish.*, 95(11) (1997) 46–52.

33. M.D. Ger and B.J. Hwang, 'Effect of surfactants on co-deposition of PTFE particles with elctroless Ni-P coating', *Mater. Chem. Phys.*, 76 (2002) 38–45.

34. P.R. Ebdon, 'The performance of electroless nickel-PTFE composites', *Plat. Surf. Finish.*, 75(9) (1988) 65–68.

35. A. Ramalho and J.C. Miranda, 'Friction and wear of electroless NiP and NiP+PTFE coatings', *Wear*, 259 (2005) 828–834.

36. E. Lanzoni, C. Martini, O. Ruggeri, R. Bertoncello and A. Glisenti, *'European Fedration of corrosion'*, Publications No. 20, The Institute of Metals, London, pp. 232–243.

37. D. Dong, X.H. Chen, W.T. Xiao, G.B. Yang and P.Y. Zhang, 'Preparation and properties of electroless Ni–P–SiO$_2$ composite coatings', *Appl. Surf. Sci.*, 255 (2009) 7051–7055.

38. T. Rabizadeh and S.R. Allahkaram, 'Corrosion resistance enhancement of Ni–P electroless coatings by incorporation of nano-SiO_2 particles', *Mater. Des.*, 32 (2011) 133–138.

39. Q. Zhao, C. Liu, X. Su, S. Zhang, W. Song, S. Wang, G. Ning, J. Yeb, Y. Linb and W. Gong, 'Antibacterial characteristics of electroless plating Ni-P-TiO_2 coatings', *Appl. Surf. Sci.*, 274 (2013) 101–104.

40. J. Novakovic, P. Vassiliou, K.L. Samara and T. Argyropoulos, 'Electroless NiP-TiO_2 composite coatings: Their production and properties', *Th. Surf. Coat. Technol.*, 201 (2006) 895–901.

41. I. Apachitei, F.D. Tichelaar, J. Duszczyk and L. Katgerman, 'The effect of heat treatment on the structure and abrasive wear resistance of autocatalytic Ni-P and Ni-P-SiC coating', *Surf. Coat. Technol.*, 149 (2002) 263–278.

42. G. Jiaqiang, L. Lei, W. Yating, S. Bin and H. Wenbin, 'Electroless Ni–P–SiC composite coatings with superfine particles', *Surf. Coat. Technol.*, 200 (2006) 5836–5842.

43. Y.S. Huang, X.T. Zeng, I. Annergren and F.M. Liu, 'Development of electroless NiP–PTFE–SiC composite coating', *Surf. Coat. Technol.*, 167 (2003) 207–211.

44. R.C. Agarwala, 'Electroless Ni–P–ferrite composite coatings for microwave applications', *PRAMANA- J Phys.*, 65 (5) (2005) 1.

45. J.T. WinowlinJappes, B. Ramamoorthy and P. Kesavan Nair, 'Novel approaches on the study of wear performance of electroless Ni–P/diamond composite deposits', *J. Mater. Process. Technol.*, 209 (2009)1004–1010.

46. R.N. Duncan, 'Hardness and wear resistance of electroless nickel- teflon composite coatings', *Met. Finish.*, 87(9) (1989) 33–34.

47. H.M. Jin, S.H. Jiang and L.N. Zhang, 'Microstructure and corrosion behavior of electroless deposited Ni–P/CeO2 coating', *Chin. Chem. Lett.*, 19 (2008) 1367–1370.

48. A.A. Zuleta, O.A. Galvis, J.G. Castano, F. Echeverria, F.J. Bolivar, M.P. Hierro and F.J. Perez-Trujillo, 'Preparation and characterization of electroless Ni–P–Fe_3O_4 composite coatings and evaluation of its high temperature oxidation behaviour', *Surf. Coat. Technol.*, 203 (2009) 3569–3578.

49. S.M.A. Shibli, B. Jabeera and R.I. Anupama, 'Development of ZnO incorporated composite Ni–ZnO–P alloy coating', *Surf. Coat. Technol.*, 200 (2006) 3903–3906.

50. P. Tao, M. Mehua, X. Feibo and X. Xinquan, 'XPS and AES investigation of nanometer composite coatings of Ni-P-ZnX on steel surface (ZnX = $ZnSnO_3$, $Zn_3(PO_4)_2$, $ZnSiO_3$)', *Appl. Surf. Sci.*, 181 (2001) 191.

51. G.O. Mallory, 'Influence of the electroless plating bath on the corrosion resistance of the deposits', *Plating*, 61 (1974) 1005.

52. B. Bozzini, V.E. Sidorov, A.S. Dovgopol and J.P. Birukov, 'Magnetic susceptibility of electroless $Ni_{84}P_{16}$ and $Ni_{84}P_{16}$-B_4C', *Int. J. Inorg. Mater.*, 2 (2000) 437–442.

53. W.X. Chen, J.P. Tu, Z.D. Xu, W.L. Chen, X.B. Zhang and D.H. Cheng, 'Tribological properties of Ni-P-multi-walled carbon nanotubes electroless composite coating', *Mater. Lett.*, 57 (2003) 1256–1260.

54. Z. Yang, H. Xu, Y.L. Shi, M.K. Li, Y. Huang and H.L. Li, 'The fabrication and corrosion behavior of electroless Ni–P-carbon nanotube composite coatings Original Research Article', *Mater. Res. Bull.*, 40 (2005) 1001–1009.

55. L.M. Ang, T.S.A. Hor, G.Q. Xu, C. Tung, S. Zhao and J.L.S. Wang, 'Electroless plating of metals onto carbon nanotubes activated by a single-step activation method', *Chem. Mater.*, 11 (1999) 2115–2118.

56. T.K. Tsai, C.C. Chuang, C.G. Chao and W.L. Liu, 'Growth and field emission of carbon nanofibers on electroless Ni–P alloy catalyst', *Diam. Relat. Mater.*, 12 (2003) 1453–1459.

57. K.G. Satyanarayana, R.C. Agarwala, V. Agarwala and S.B. Sharma, 'Patent for -Ni-P-Al_2O_3-ZrO_2 EL Composite Nanocoating', Filed through RRL-Thiruvananthapuram, CSIR Lab: India, Gazette No. 100/Del/2001.

58. (a) S. Maheswary, S. Sharma and A. Sharma, 'Corrosion investigation on conventional and nanocomposite (Ni-P-Al$_2$O$_3$-TiO$_2$) coated mild steel by in-plant test in digester of a paper mill', *Ind. J. Sci. Technol.*, 9(33) (2016) 1–8. (b) R. Kumar, S. Sharma and A. Sharma, 'Corrosion study of electroless Ni-P-Al$_2$O$_3$-ZrO$_2$ nanocomposite coatings in paper mill digester', *Ind. J. Sci. Technol.*, 9 (44) (2016) 1–8.
59. SMA Shibli, VR Anupama, PS Arun, P Jineesh and L. Suji, 'Synthesis and development of nano WO3 catalyst incorporated Ni–P coating for electrocatalytic hydrogen evolution reaction', *Int. J. Hydrog. Energy*, 41 (24) (2016).
60. C. Wang, Z Farhat, G. Jarjoura, M.K. Hassan and A.M. Abdullah, 'Indentation and bending behavior of electroless Ni-P-Ti composite coatings on pipeline steel', *Surf. Coat. Technol.*, 334 (25) (2018) 243–252.
61. Z. Li, Z. Farhat, G. Jarjoura, E. Fayyad, A. Abdullah and M. Hassan, 'Synthesis and characterization of scratch-resistant Ni-P-Ti-based composite coating', *Tribol. Trans.*, 62 (2019) 880–896.
62. H. Luo, M. Leitch, H. Zeng and J. Li Luo, 'Characterization of microstructure and properties of electroless duplex Ni-W-P/Ni-P nano-ZrO$_2$ composite coating', *Mater. Today Phys.*, 4 (2018) 36–42.
63. D.R. Dhakal, G. Gyawali, Y.K. Kshetri, J. Hyuk Choi, S.W. Lee, 'Microstructural and electrochemical corrosion properties of electroless Ni-P-TaC composite coating', *Surf. Coat. Technol.*, 381 (2020) 125–135.

Index